QUESTIONS & ANSWERS

Colour Television

J. A. Reddihough

and

David Knight

Newnes Technical Books

THE BUTTERWORTH GROUP

ENGLAND
Butterworth & Co. (Publishers) Ltd
London: 88 Kingsway, WC2B 6AB

AUSTRALIA
Butterworths Pty Ltd
Sydney: 586 Pacific Highway, NSW 2067
also at Melbourne, Brisbane, Adelaide and Perth

CANADA
Butterworth & Co. (Canada) Ltd
Toronto: 2265 Midland Avenue, Scarborough, Ontario M1P 4S1

NEW ZEALAND
Butterworths of New Zealand Ltd
Wellington: 26–28 Waring Taylor Street, 1

SOUTH AFRICA
Butterworth & Co. (South Africa) (Pty) Ltd
Durban: 152–154 Gale Street

USA
Butterworth (Publishers) Inc
Boston: 19 Cummings Park, Woburn, Mass. 01801

First published in 1969 by Newnes-Butterworths
Second edition by Newnes Technical Books,
a Butterworth imprint, 1975
Second impression 1976

ISBN 0 408 00162 3

Printed and bound in England by
Hazell Watson & Viney Ltd, Aylesbury, Bucks

CONTENTS

PREFACE

The aim of this book is to provide a simple, practical account of colour television transmission and reception. The emphasis has been placed on receiver techniques since it is these that will be of most concern to the enthusiast, technician and service engineer.

The book is based on the PAL system throughout. An elementary knowledge of radio and television has had to be assumed in order to keep the book within the limits of the series. Those wishing to delve first into radio and television generally should read *Q and A on Radio and Television* by H. W. Hellyer.

In updating this popular little book I have taken into account most of the new developments that have occurred in colour receivers since the first edition was written. I have endeavoured not to deviate from the simple, practical approach carefully established by John Reddihough, the original author, and the four sections have been retained. I have introduced a number of new illustrations and have included new information on such things as display tubes, 110° tubes and circuitry, RGB primary colour drive to the picture tube, the use of integrated circuits and on some of the more recent chrominance circuitry. A number of new questions and answers have been included and some of the original ones have been brought up to date.

I am indebted to John Reddihough for his help in establishing the new material to be included and for his detailed list of references, without which the updating would have proved a much more protractive experience.   D.K.

COLOUR SIGNALS
AND TRANSMISSION

What is light?

Light is a form of electromagnetic radiation, just as are heat, radio transmissions, X- and gamma-rays and so on. The frequencies of light radiation lie in the band between 385×10^6 and 790×10^6 MHz. As shown in Fig. 1, light frequencies occupy the portion of the electromagnetic wave spectrum between infra-red and ultra-violet radiation.

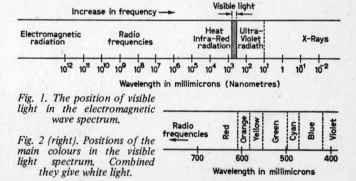

Fig. 1. The position of visible light in the electromagnetic wave spectrum.

Fig. 2 (right). Positions of the main colours in the visible light spectrum. Combined they give white light.

What is the difference between coloured and white light?

White light, as experiments with prisms show, is a combination of all the various colours. If the portion of the electromagnetic wave spectrum representing light—or visible—radiation is examined, it is found that this extends from blue light at approximately 400 mμ (millimicrons) to red light at approximately 700 mμ (see Fig. 2). A millimicron (SI unit, millimicrometre) is a thousand-millionth of a metre, i.e. a nanometre.

How are the colours related to one another?

The relationship between the various colours—and white light—can be shown by means of the chromaticity diagram (see Fig. 3). In this the colours, as they merge into one another, are plotted around the edge of the diagram. This is of course, as shown, a plot in terms of the wavelength of the various colours. Thus the dominant wavelength of green light is about 520 mμ: at lower wavelengths green light merges, via cyan, into blue light, whilst at higher wavelengths it merges via yellow into red light. No dominant frequency can be given for

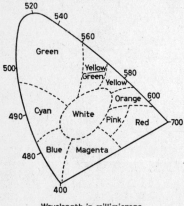

Fig. 3. The chromaticity diagram.

Wavelength in millimicrons

the shades between blue and red: these shades, the purples, are a combination of red and blue radiation with complete absence of the green section of the colour spectrum. White light, being a combination of all the colours, occupies the centre of the chromaticity diagram.

What are the "primary colours"?

In radiated light, red, green and blue. As can be seen from Fig. 3 they occupy roughly the corners of the chromaticity diagram. Their importance lies in the fact that all other colours can be obtained from combinations in various proportions of

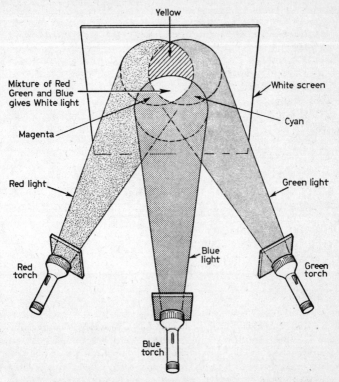

Fig. 4. A combination of the three primary colours red, blue and green gives white light. In this simple experiment three torches with filters in front produce red, blue and green beams of light which where they overlap on the white screen produce white illumination.

any two primary colours. And, as shown in Fig. 4, combining the three primary colours gives white light.

How does this make colour television possible?

Because the scene to be televised can easily be analysed in terms of the amounts of the three primary colours present.

Why do these primary colours differ from those used in painting?

The primary colours used in painting are red, yellow and blue. The difference lies in the fact that in looking at a painted surface we see reflected light instead of radiated light. The characteristics of reflected and radiated light (e.g. sunlight or light from a source such as an electric light bulb) differ. In the case of radiated light the primary colours being radiated *add* together to give different colours: the colour mixing process is said to be *additive*. In the case of light being reflected from a coloured surface, however, what happens is that the surface absorbs all the light reaching it except the particular colour which it reflects; that is, in this case the colour mixing is *subtractive*.

How is the scene to be televised analysed in terms of the amounts of the primary colours present?

If the television camera is fitted with three pick-up tubes (the camera equivalent of a receiver's cathode-ray or picture tube) one can be used to record the amount of each primary colour present. A simple system of dichroic lenses and mirrors can, as shown in Fig. 5, be used to separate the primary colours and present them to each pick-up tube separately, a colour filter in front of each tube being used to adjust the colour balance between the three tubes. In this way three electrical signals are obtained from the three pick-up tubes, one representing the amount of red light present, a second the amount of green light present and the third the amount of blue light present.

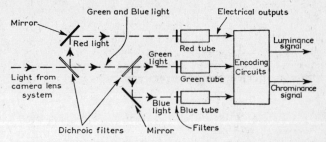

Fig. 5. How in a three-tube colour television camera an optical system with two dichroic filters separates the light inputs to the three tubes which then provide outputs representing the amount of red, green and blue light present. Dichroic filters allow light at certain frequencies to pass straight through while reflecting light at other frequencies. They can thus as shown split the incoming light into the three primary colour components. The filters provide additional colour balancing between the tubes. In practice the signals are gamma corrected before being encoded. Gamma correction compensates for the fact that whereas the output from a camera tube is linear a receiver c.r.t. operates from cut-off (black) to white, dark tones occurring in a very non-linear part of the c.r.t.'s characteristic.

How can a particular shade of colour be defined?

Apart from its brightness, i.e. the strength of the illumination present, which is what monochrome television and films record, the colours we see can be defined in terms of their *hue* and *saturation*.

What is meant by hue and saturation?

Hue represents the dominant wavelength of a colour. Referring back to the chromaticity diagram, Fig. 3, we see that the various colours in their pure form are plotted around the edge of this diagram in the order of their wavelength. At the centre of the diagram the colours merge together to give white light. Thus as we move from the outside of the diagram towards the centre, the hues become gradually less vivid, merging in the centre into white light. The purity of a particular hue, i.e. where it falls between its pure state at one extreme and white light at the other, is termed its degree of saturation.

To summarise: the three characteristics of coloured light are (a) its intensity, or brightness—referred to in colour television as its *luminance*— which is independent of the particular colour, (b) its hue, i.e. dominant wavelength, and (c) its saturation or purity, i.e. the extent to which it is mixed with light of other wavelengths, which reduce the vividness of the particular hue.

In what ways do colour television cameras differ from black-and-white only cameras?

A camera for *monochrome* (i.e. black-and-white only) television records only the average brightness level of the light present, for which a single camera pick-up tube is needed. In a colour television camera, however, three camera tubes are required, as shown in Fig. 5, to record the amount of each primary colour in the scene in front of the camera. In addition, in some colour television cameras a fourth pick-up tube is used to record the brightness of the scene before the camera, as in monochrome television.

Why is this necessary?

Because colour television transmissions must be *compatible*, that is capable of being received by black-and-white only receivers, which are able to receive the luminance information from which to build up a monochrome picture but do not respond to the transmitted colour signals. Since at the start of colour transmitting the vast majority of sets were monochrome ones it was clearly necessary for the colour service, which must use the same channels and frequencies as the existing black-and-white only service, to be compatible.

How is a compatible signal obtained where a three-tube colour camera is used?

Since the three tubes—R (red), G (green) and B (blue)—between them record the luminance of the scene before them, a luminance signal, called the Y signal, can be obtained by adding their outputs together. The human eye, however, is not equally sensitive over the colour spectrum, i.e. it has greater sensitivity to some colours than to others (its maximum sensitivity is to green-yellow shades), and for this reason it is necessary to add the R, G and B signals together in certain proportions to obtain a luminance signal that corresponds to the characteristics of the human eye. These proportions are 0·3R, 0·59G and 0·11B. Thus

$$Y = 0·3R + 0·59G + 0·11B.$$

Some of the more recent colour television cameras feature a fourth tube essentially for producing the luminance or Y signal, in addition to the three tubes for the red, green and blue primary colour signals. There have also been developments in the production of single-tube colour cameras: by using a striped optical filter on the faceplate a complex output which can be separated by electronic filters is generated.

It is noteworthy that the tubes used in all present day cameras for broadcasting (three- and four-tube types) are lead oxide vidicons (Plumbicons), which are of the photoconductive type.

How many signals are there in a colour television transmission?

It may at first appear that four signals are required, Y, R, B and G. However, since the Y signal represents the total R, B and G light brightness present, it is only necessary to transmit the Y signal, which is required for compatible transmission, and two colour signals, since the third colour will be the amount of light present that is not accounted for by the two colour signals actually transmitted. In practice the Y signal is transmitted along with two *colour-difference* signals, R−Y and B−Y. In the receiver it is a simple matter to derive the third colour-difference signal, G−Y, by adding together suitable proportions of −(R−Y) and −(B−Y). A simple resistor matrix network (see Section 3) performs this operation.

What is a colour-difference signal?

The colour-difference signals are formed by subtracting the luminance signal from each of the primary colour signals. Since the Y signal consists of 30 per cent of the red light present, 59 per cent of the green light present and 11 per cent of the blue light present, the colour information transmitted must indicate to the receiver the amount of red or blue light present (information on green not being transmitted as we have seen) that is not present in the Y signal. To take an example: when a pure red (i.e. fully-saturated red) part of the scene is being scanned by the camera a red signal and a Y signal that is 30 per cent of the red signal will be produced. Since the receiver must receive the Y signal the information it will need in *addition* to produce the original red is the *difference* between, in this example, the Y signal and the red signal produced by the R tube in the camera, that is R−Y. In this way adding together the Y information (30 per cent R) and the red colour-difference information (R−Y=70 per cent R) gives us the original R signal.

What is meant by encoding?

Encoding is the process of obtaining from the camera tube outputs the Y, B−Y and R−Y signals that are transmitted. As shown in Fig. 5 the camera tube outputs (a three-tube camera being assumed in this example) are fed to an encoding unit. The output from this is the Y (luminance) and the two colour-difference signals which are together known as the *chroma* signal. In a colour receiver a *decoder* is needed to separate the various components of the colour television transmission.

How are the colour-difference signals obtained?

The encoding process is illustrated in Fig. 6. First the R, G and B outputs from the camera tubes are added together in the correct proportions to give the Y signal. The other two outputs, the colour-difference signals, are obtained by feeding the Y signal via a phase inverter to give the required −Y signal to two further adders which provide the R−Y and B−Y colour-difference signals.

How are the colour-difference signals used?

We have already given the example of a fully saturated red signal. Expressing this in terms of voltages, suppose that our fully-saturated red signal, the output from the red camera tube

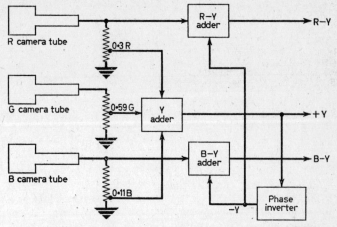

Fig. 6. The encoding process: how the Y, R–Y and B–Y signals are obtained using a three-tube camera.

in Fig. 6, is 1 V. Thus we have 0·3 V fed to the Y adder and this, since there are no blue or green outputs, is the Y signal. By inverting this we obtain −0·3 V and this, added to the 1 V red signal, gives us an R–Y signal of 0·7 V. When the two signals, Y 0·3 V and R–Y 0·7 V, are applied to the receiver and added together, the result is the recovery of the original 1 V red signal.

How are the true colour signals derived from the colour-difference signals in the receiver?

The type of colour television tube in general use is the three gun shadowmask tube developed by RCA. In this three beams—one for each primary colour—scan out the picture together, one beam being emitted by each of the three R, G and B guns in the tube. Each beam activates a different type of colour phosphor on the screen of the tube thus giving a full-colour picture. If the Y signal is fed, as shown in Fig. 7, to each of the three cathodes simultaneously, while the R–Y, G–Y and B–Y colour-difference signals are fed to the three

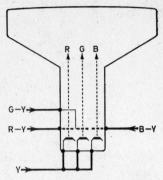

Fig. 7. Modulating the three electron beams in a three-gun colour tube to obtain true R. G and B signals. The luminance signal is fed to the three cathodes simultaneously and the three colour-difference signals applied to the respective grids.

control grids, the result is that the picture tube recovers the three true R, G and B signals which then activate the phosphor-coated tube screen. In other words, with Y at the cathodes and R−Y, B−Y and G−Y at the control grids we have R, G and B at the screen.

Reverting to our example of the fully saturated red signal, we have seen that adding the 0·3V Y signal to the 0·7V R−Y colour-difference signal will give us the original 1V red signal. This occurs at the red gun. Now the 0·3V Y signal is also fed to the blue and green guns but these will also receive −0·3V colour-difference signals since, there being no blue or green light present, the B−Y and G−Y colour-difference signals represent 0V − 0.3V in each case. Thus in spite of the fact that the Y signal is fed to all three cathodes simultaneously the result, in our example, is that two guns will be cut off whilst the other produces our 1V fully saturated red signal.

In the arrangement illustrated in Fig. 7 the shadowmask tube acts as the final matrix to recover the three primary colours. Alternative methods of matrixing can be employed, for example using resistor matrix networks in the same way as is done in establishing the G−Y signal.

What is meant by RGB drive?

Since the first edition of this book was written there has been an almost general change in receiver design so instead of

16

the true colour signals (or primary colour signals as they are often called) being derived at the guns of the picture tube for individual beam modulation they are derived in stages prior to the picture tube and then applied to the guns (usually to the cathodes) as primary colour signal drive. The primary colours are red, green and blue, hence the term RGB drive.

In simplest form, each primary colour is obtained from a circuit which effectively adds the Y signal to the colour-difference signal, so that on the red channel, for example, we get $(R-Y) + Y=R$. When a transistor is used for this algebraic addition the Y signal may be applied to the base and the colour-difference signal to the emitter, so that at the collector we get the primary colour signal. As this may be of insufficient amplitude fully to drive the appropriate gun each channel may contain a primary colour amplifier (sometimes termed "colour video") between the primary colour matrixing stage and the cathode of the appropriate gun.

Integrated circuits are also featuring in this area of the contemporary colour receiver, and various techniques have been evolved to provide brightness control and black level clamping to avoid changes in key brightness of the display with changes in picture content.

How is the G−Y signal recreated in the receiver?

Since the luminance signal Y represents the amount of red, blue and green light present (in certain proportions as we have seen) and the B−Y and R−Y signals represent the difference between the red and blue signals and the luminance, Y, signal, it can be shown that adding together $-(B-Y)$ and $-(R-Y)$, signals which are simply obtained by inverting the phase of the demodulated B−Y and R−Y signals, gives the G−Y colour-difference signal. This addition has to be done in proportions that correspond to the proportions in which the R, G and B signals were originally added together to give the Y signal. The proportions are:

$$G-Y= -0.51 (R-Y) -0.19 (B-Y).$$

The addition is done by a simple resistor network the values of which are selected to give the above proportions. The same general technique is used when the design features RGB drive.

What is meant by gamma correction?

The colour signals picked up by the camera tubes are altered slightly before being encoded to take into account the different characteristics of camera tubes as opposed to receiver colour display tubes. This adjustment, which is also necessary in monochrome television, is termed gamma correction, and signals that have been given this correction are usually indicated as R',G',Y', etc.

What is meant by weighting the B−Y and R−Y colour-difference signals?

The application of the full R−Y and B−Y colour-difference signals to the modulators in the transmitter would result in overmodulation and for this reason the amplitude of the R−Y and B−Y colour-difference signals, is reduced or weighted as it is called, prior to modulation. The R−Y signal is reduced to 0·877 (R−Y) and the B−Y signal to 0·493 (B−Y). The signals are equalised in the receiver by adjusting the gains of the R−Y and B−Y signal amplifiers. See under V and U signals, p.28.

What bandwidth is required for colour television transmission?

Because of the limited band space available it is desirable to keep to the same channel spacings, i.e. bandwidths, used for black-and-white television. Three factors make this possible.

First, the human eye is not very sensitive to colour detail, so that less colour information than information on black-and-white detail, that is changes in light intensity, needs to be transmitted. In fact compared with the 6·75 MHz vestigial sideband vision signal bandwidth required for the transmission of a 625-line monochrome picture, the bandwidth required by the chroma signal to give good results is approximately ±1 MHz.

Secondly, the two colour-difference signals actually sent out are transmitted by two-phase modulation using a single chroma subcarrier frequency—a system called quadrature modulation. The principle is similar to the way in which dual-channel stereo sound can be modulated on to a single f.m. carrier wave.

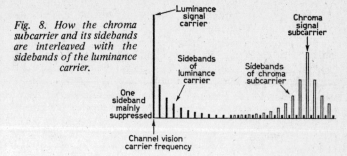

Fig. 8. How the chroma subcarrier and its sidebands are interleaved with the sidebands of the luminance carrier.

Thirdly, because of the line structure of television pictures there are sections of the black-and-white channel bandwidth in which no information is carried. If the chroma subcarrier frequency is made an odd multiple of half the line frequency the chroma signals can be interleaved with the luminance information, occupying those portions where luminance information is not present, giving a channel bandwidth of 8 MHz as with monochrome transmission. The principle is illustrated in Fig. 8: the luminance signal carrier is shown on the left in solid black, with its sidebands that carry the luminance information—also shown in solid black—extending across to the right (remember that the sidebands to one side of the carrier are partially suppressed in the vestigial sideband system used for black-and-white television transmission). Because of the line structure of the picture, gaps appear as shown between the luminance carrier sidebands. The chroma subcarrier is shown, with its sidebands, occupying these gaps in the luminance signal sideband structure. The subcarrier frequency must be very closely controlled to make this technique effective: its frequency is actually 4·43361875 MHz, which indicates the order of accuracy required (but to save space we shall subsequently refer to it as 4·43 MHz).

How is the chroma subcarrier modulated?

The technique used is to start with two subcarriers at the same frequency but with a phase difference between them of

19

90°. One is amplitude modulated by the R−Y colour-difference signal, the other being amplitude modulated by the B−Y colour-difference signal. The two 90° out-of-phase individually amplitude modulated subcarriers are then added together, giving a single "two-phase" or quadrature-modulated subcarrier. This quadrature-modulated subcarrier is then amplified, along with the luminance signal, in the transmitter's r.f. amplifier chain before being fed to the transmitting aerial. In both the U.S. NTSC and the PAL systems suppressed subcarrier modulation is used so that at the output of the R−Y and B−Y modulators the two subcarriers—at the same frequency but with 90° phase difference between them—are cancelled out but the modulation products—the sidebands of the subcarriers, which carry the actual chroma information—appear and are passed on to the transmitter's r.f. amplifier chain via the adding network. The basic scheme is shown in block diagram form in Fig. 9. To obtain suppressed subcarrier modulation a balanced modulator is used.

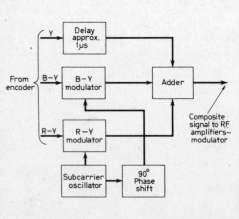

Fig. 9. Basic scheme for two-phase modulation of the chroma subcarrier. The delay line in the luminance channel keeps the Y signal in step with the B−Y and R−Y signals as they pass through the modulator stages. In practice for reasons given later the signal from the subcarrier oscillator is passed through an inverter (180° phase shift) on alternate lines in the PAL system before being applied to the R−Y modulator.

What is the effect of adding together the modulation products of the two 90° out-of-phase subcarriers?

The resultant obtained by adding together the modulation products of the two 90° out-of-phase (suppressed) subcarriers is a signal that is the vector sum of the modulation products of the two subcarriers. This vector sum resultant is a signal that varies *in both amplitude and phase*. The principle is illustrated, in somewhat simplified form, in Fig. 10. Here (a) shows the two subcarriers, R−Y and B−Y, with 90° phase difference between them and, in this instance, of equal amplitude. The resultant, or vector sum, is found by completing the square, as shown by the dotted lines, and drawing the diagonal line from the point denoting nil modulation of either subcarrier to the dotted corner of the square. This diagonal, then, is the resultant obtained by adding together the R−Y and B−Y subcarriers. At (b) and (c) are shown first the case where the R−Y subcarrier is four times the amplitude of the B−Y subcarrier and secondly the case where the R−Y subcarrier is a quarter of the amplitude of the B−Y subcarrier. These examples are given to show how the resultant varies in both amplitude and phase according to the amplitude of the two subcarriers.

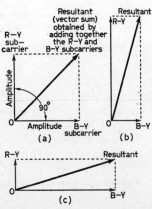

Fig. 10. How two signals 90° apart in phase when added give a vector sum or resultant that indicates the combined amplitude and phase characteristics of the two signals.

How does this enable transmission of information on hue and saturation to be achieved?

With two subcarriers 90° out-of-phase, one amplitude modulated to indicate the amount of red light saturation, if any, present in the light of the scene being televised at a given time along the line being scanned, and the other amplitude modulated to indicate the amount of blue light saturation, if any, at the same time and point, the resultant obtained by adding these subcarriers together is a signal that indicates by its phase the dominant hue of the original light and by its amplitude the saturation of the original light. Fig. 11(a) illustrates this. In Fig. 11(b) the two subcarriers are shown superimposed on the chromaticity diagram (which has had to be moved around for this purpose). From this it can be seen that nil modulation of both subcarriers represents white light (no colour present) while all possibilities between a fully-saturated red hue and a fully-saturated blue hue can be denoted by various resultants within the quadrant A—B obtained by adding together the red and blue subcarriers in various proportions, i.e. by means of signals such as those shown in Fig. 11(c).

Although the subcarrier is suppressed at the transmitter, as we have already seen, there remains a signal based on the

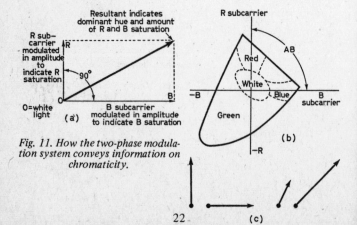

Fig. 11. How the two-phase modulation system conveys information on chromaticity.

22

subcarrier frequency, of course, of amplitude dependent on that of the phasor (Fig. 11).

How does the receiver resolve two-phase modulation?

Since in the transmitter two modulators 90° apart in phase are used to produce the transmitted chroma signal, for the receiver to resolve the received signal so as to recreate the two separate colour-difference signals it requires two demodulators operating with a 90° phase difference. How this works out can be seen from Fig. 12, taking again as our example a fully saturated red part of the scene being televised. As we have seen in the case of a 1 V output from the red camera tube this would give rise to colour-difference signals of 0·7 V in the case of the R−Y signal and −0·3 V in the case of the B−Y signal. This provides the resultant shown in Fig. 12. If in the receiver we have two demodulators operating 90° out-of-phase, one sampling the signal on the R−Y axis and the other sampling the signal on the B−Y axis, the R−Y demodulator will "see" an 0·7 V signal while the B−Y demodulator "sees" a −0·3 V signal. Thus the original two colour-difference signals are recovered from the transmitted resultant or vector sum (i.e. the two-phase modulated) signal. The type of demodulators required are synchronous, or "switched" demodulators, instead of simple envelope detectors (see pages 81-84). A synchronous demodulator can give either a positive or a negative output.

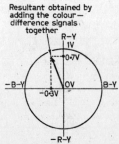

Resultant obtained by adding the colour—difference signals together

Fig. 12. B−Y and R−Y colour-difference signals for a fully saturated red area of the scene being scanned by the camera. The resultant obtained by two-phase modulation is resolved in the receiver into the original 0·7V and −0·3V components by demodulators which operate on the R−Y and B−Y axes. How these synchronous demodulators, as they are called, work is explained in Section 3.

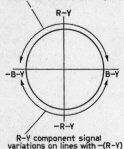

R−Y component signal
variations on lines with +(R−Y)

R−Y

−B−Y   B−Y

−R−Y

R−Y component signal
variations on lines with −(R−Y)

Fig. 13. In the PAL system the R−Y component of the chroma signal is inverted on alternate lines. For example, a component of, say, +(R−Y) becomes −(R−Y) on a phase-inverted line. How this action compensates for spurious phase shifts, thereby combating colour errors, is explained by Fig. 14.

Are there any problems associated with the two-phase modulation technique?

The main problem is that signals passing through electronic equipment can be subject to spurious phase changes. This means that unless some compensatory technique is used incorrect colours may be received and displayed by a colour television receiver.

What methods can be used to overcome this problem?

In the U.S. NTSC system a hue control is provided which can be adjusted by the viewer to correct phase changes giving rise to incorrect colouring. In the PAL system, which was developed to overcome this particular problem, the compensation required to correct the effects of spurious phase shifts is automatically provided. PAL is an abbreviation for Phase Alternation Line: what this means is that on alternate lines the phase of one of the colour-difference signals, in practice the R−Y signal, is reversed, i.e. it is transmitted on alternate lines with a phase change of $180°$. Say on one line it is transmitted as R−Y, on the next it will be transmitted as −(R−Y). This is illustrated in Fig. 13. On one line R−Y signal variations occur, as shown, in the area at the top of the diagram, whilst on the succeeding line with $180°$ R−Y signal phase inversion R−Y singal variations, now −(R−Y), occur in the area at the bottom of the diagram. This assumes of course that the R−Y

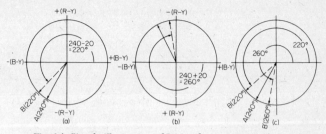

Fig. 14. Simple illustration of PAL colour correction by averaging.

signal itself does not change from positive to negative values between lines.

How does this solve the problem of spurious phase shift?

In Fig. 14(a) full-line phasor A corresponds to a "greenish" signal component of $240°$ phase, but owing to an assumed $20°$ phase error this appears as broken-line phasor B at $220°$. Diagram (b) shows the effect of the phase reversal on the next line of signal component, where phasor B now has an effective phase of $260°$. Thus, on the non-reversed line phase B is at $240-20=220°$, while on the reversed line it is at $240+20=260°$.

Diagram (c) gives an elementary impression of how the PAL receiver automatically compensates for such a phase error, it merely averaging the non-reversed and phase-reversed errors, but at the expense of mild desaturation in the PAL-D system (see later). In the example, the average phase of $220°$ and $260°$ is $240°$, which corresponds to the phasor bearing of the signal component of correct phase.

The action is more complicated than implied by simplified diagram (c) for in a PAL-D receiver a delay line is used so that direct and delayed signals can be added and subtracted to provide the averaging electronically, and account then has to be taken of the $180°$ phase shift given to the delayed signal by the line.

Without this sort of correction it would be necessary to arrange for manual phase adjustment (by the American hue control, for example) to allow the viewer to secure the correct hue display from signal components in phase error. Uncorrected, an error as indicated by Fig. 14 would tend to cause a green component to veer towards yellow.

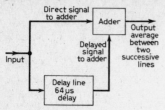

Fig. 15. How a delay line and adder unit are used to carry out the averaging process between the signals on alternate lines in the PAL system.

How is the averaging process carried out?

The simplest method is optical averaging: as the eye is not very sensitive to colour detail the effect of displaying alternate lines of the picture with compensating errors would be to give an average effect of the true colour. Improved results, however, are obtained by the use of electronic averaging. To do this a delay line having a delay time of exactly one line duration (approximately 64μs for a 625-line picture) is required in the receiver. This delay line enables the received colour information at the same point along succeeding pairs of lines to be compared to obtain the average result. The principle is outlined in Fig. 15. The chroma signal received is fed directly to the adder unit and also to the adder via the one-line duration delay line. The output of the delay line being the chroma signal at the same point along the preceding line, the adder is able to average the chroma information along successive pairs of lines.

Are there any problems associated with the use of suppressed subcarrier transmission of the chroma signal?

The only complication this introduces is that in order to recover the chroma information in the receiver it is necessary for the receiver to include a stable reference oscillator that operates at the same frequency and in phase with the transmitter subcarrier oscillator. This oscillator controls the two (B−Y and R−Y) demodulators in the receiver.

How are the transmitter subcarrier and receiver reference oscillators synchronised?

A "burst signal" comprising 10 Hz (±1 Hz) of the subcarrier frequency is transmitted during the back porch period of the television waveform, that is immediately following the line sync pulse. This burst signal is used in the receiver to synchronise the reference oscillator by means of an automatic frequency control circuit. The average phase of the transmitted bursts is along the −(B−Y) axis.

Because of the phase reversal of the R−Y component of the chroma signal on alternate lines, the burst signal varies in phase by ±45° on alternate lines. It is +45° when R−Y is not inverted, and −45° when the line is transmitted with inverted R−Y chroma information. This enables the receiver to distinguish between the lines inverted and non-inverted with R−Y components. The reference oscillator locks in quadrature with the average phase of the bursts.

What is Illuminant D?

As anyone who has used several cans of white paint of different makes knows, no two are the same! Because with radiated white light there is an amalgam of light of all colours, there is obviously the possibility of various "whites" here, too − the differences occurring in the white area of Fig. 3. This can be a problem to the service technician, and to obtain the correct white light he needs a reference white to make the appropriate receiver adjustments (grey scale, etc.), for which various aids are available to him.

The white now adopted for colour television is called Illuminant D, which has a colour temperature of 6500 K. The concept of colour temperature is based on the temperature to which a black body − one which absorbs all light − needs to be raised to radiate a given white light.

Illuminant D simulates "standard daylight" which comprises a mixture of direct sunlight and diffuse skylight. It lies slightly towards the green side of the black body locus, while the bias of the previously used Illuminant C is slightly towards the magenta side.

Certain preset controls in the receiver can be adjusted to change the white light, hence the need for a reference. At the transmitting end, of course, it is not possible to ensure that all lighting conforms to Illuminant D, but before transmission the signal is adjusted to give the effect of originating under Illuminant D conditions.

What are U and V signals?

These are alternative ways of referring to the B−Y and R−Y signals respectively but indicate that the signals have been weighted as described earlier.

What type of aerial is required for colour reception?

Since transmission is on the same channels and with the same bandwidth as black-and-white transmissions, there are no special aerial requirements and an aerial that gives a good, clean signal on monochrome will be satisfactory for colour. The important words here are good and clean (i.e. free of grain): to give satisfactory colour reception the aerial system should have some reserve of signal, and an aerial that is just satisfactory on black-and-white will not do justice to the colour transmissions. The signal provided by the aerial system for the best colour reception should be a minimum of 1·5 mV on Band IV rising to a minimum of 2·25 mV on Band V. The need for a little extra signal for colour reception arises from the fact that the colour-difference signals on pale colours are of very low amplitude—on white, black and grey they are not present at all, but appear at low strength with light pastel shades. If an existing monochrome aerial system has some reserve of signal (which will be cancelled on black-and-white by the set's a.g.c. system), is correctly matched to the receiver and uses low-loss feeder, then the results on colour will be satisfactory. Careful aerial installation is important, particularly to avoid signal cancellation produced by multi-path reception and other propagation effects.

Can you give a summary of the transmitting end of the PAL system, showing the main items in the chain and the signals produced by the colour bars?

This is not particularly easy, but the diagram in Fig. 16 may go some way to answer it. Here is shown a three-tube colour camera scanning a colour bar pattern (white, yellow, cyan, green, magenta, red, blue and black) and the signal waveforms so yielded. Each of the three tubes in the camera are carefully adjusted to respond separately — yet in perfect registration — to the scene being televised, one receiving light input through a red filter, the second through a green filter and the third through a blue filter. In this manner the scene is "analysed" in terms of the three additive primary colours, the correctly proportioned addition of which, as we have seen, yields white light. Thus we obtain separate red, green and blue primary colour signals.

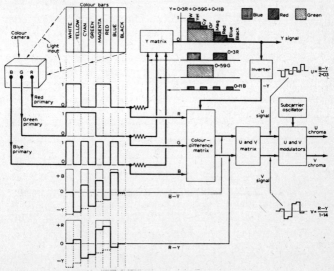

Fig. 16. Diagram showing the front-end of the PAL TV system and the various main signals involved (see text for more details).

It should be noted that red output occurs from the white, yellow, magenta and red bars, green from the white, yellow, cyan and green bars, and blue from the white, cyan, magenta and blue bars, the reason for which should now be understood.

After leaving the camera outputs the constant (unity) amplitude primary colour signals are adjusted in amplitude to provide 30 per cent red, 59 per cent green and 11 per cent blue and then they are added (in the Y matrix) to yield the Y signal. This signal is also adjusted in amplitude (giving the white step unity amplitude in the diagram). The stepped nature of the Y signal from the colour bars stems from the addition of the level-adjusted primary colour signals, and since the amplitude of each of these signals has been "normalised", it follows that the white step will have an amplitude of unity (i.e. $0.3 + 0.59 + 0.11 = 1$). The other steps down from white will then have amplitudes of 0·89 yellow, 0·7 cyan, 0·59 green, 0·41 magenta, 0·3 red, 0·11 blue and, of course, zero for black. This, then, is the Y signal which corresponds to the monochrome vision signal of the same scene obtained from a single-tube black-and-white camera.

The R−Y and B−Y signals for transmission are obtained by subtracting the Y signal separately from the red and blue primary colour signals, giving R−Y and B−Y. The G−Y signal is not transmitted, this being reconstructed at the receiver by G−Y matrixing as we have seen. The colour-difference matrix, which receives inverted Y signal as well as the primary colour signals, produces the two colour-difference signals, and these, after PAL weighting, are then applied to the modulators.

The modulators also receive the subcarrier signal, but as we have seen this is suppressed, only the R−Y and B−Y chroma components (commonly called V and U chroma signals after the PAL weighting) which are based on the subcarrier frequency being applied, along with the luminance signal, bursts, sync pulses, black-level intervals, etc., to the vision signal carrier wave for modulation.

2

PICTURE DISPLAYS ON
COLOUR TUBES

What type of picture tube is used to display the colour pictures in a colour television receiver?

The advent of colour television saw the exclusive use of the three-gun shadowmask tube.

Is this type of tube still used today?

Yes, it certainly is, but different tubes are coming into being and a tube other than the shadowmask may be found in a contemporary receiver. However, before we look at one or two of the more recent types of display it would be as well to see how the shadowmask tube works since there are many millions of these operating the world over.

How does the shadowmask tube differ from the normal kind of black-and-white tube?

It differs in three main respects. First there are three electron guns providing three electron beams, one for each primary colour, and to accommodate these the neck of the tube may be wider than the neck of a monochrome counterpart — 38 mm (about 1½ in) — so due to this and the greater energy carried by the electron beams (they can be regarded as "stiffer" than the beam of a monochrome tube owing to the higher e.h.t. voltage applied to the final anode) greater energy is required to deflect the beams. More modern shadowmask tubes, however, use less wide necks and a greater angle of deflection, which makes the length of the neck shorter, too, but the wider scanning angle, in spite of the less wide neck,

31

The red, green and blue phosphor dots are deposited on the screen in triad groups as shown

Fig. 17. (left). How the triad groups of red, green and blue phosphor dots on the screen of the shadowmask tube and the holes in the shadowmask are aligned. During manufacture the phosphor dots are actually deposited on the screen of the tube through the holes in the shadowmask.

The holes in the shadowmask are smaller than the phosphor dots on the screen

Part of the shadowmask

Fig. 18 (below). Principle of the three-gun shadowmask tube. The three beams converge at the shadowmask, passing through its holes and then diverging so as to fall on phosphor dots of the appropriate colour.

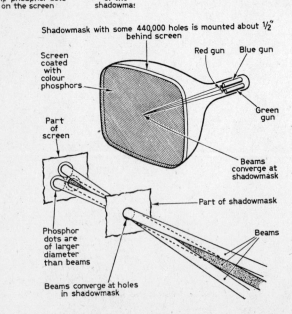

Shadowmask with some 440,000 holes is mounted about ½" behind screen

Screen coated with colour phosphors

Red gun   Blue gun

Green gun

Beams converge at shadowmask

Part of screen

Part of shadowmask

Phosphor dots are of larger diameter than beams

Beams

Beams converge at holes in shadowmask

still makes the requirement for scanning energy higher than for most monochrome tubes.

Secondly, the screen is coated with three different types of phosphor, which emit separately red, green and blue light under the electron bombardment from the three guns. The phosphors are deposited on the screen in a triangular dot pattern as shown in Fig. 17. Each group of three dots is called a triad.

Thirdly, a shadowmask — from which the tube takes its name — is accurately mounted in the flare about 12 mm (½ in) behind the screen.

What is the purpose of the shadowmask?

The shadowmask ensures that when the tube and its associated circuits are correctly set up the green beam lands only on green phosphor dots on the screen, the blue beam only on blue phosphor dots, and the red beam only on red phosphor dots. The mask has some 440,000 holes in it and each of these is aligned, as shown in Fig. 17, with one of the triads of phosphor dots on the screen. The three guns are so mounted that their beams *converge*, i.e. come together, at the shadowmask and then diverge again to activate the phosphor dots on the screen. Thus in effect three separate pictures are being scanned out, each one being produced by a different type of phosphor, the three pictures of course being super-imposed so far as the viewer is concerned so that when colour signals are applied to the tube he sees a coloured picture. For simplicity in Fig. 18 the three beams are shown converging on a single shadowmask hole: in practice the diameter of the beams is appreciably greater than that of the holes and the beams actually pass through a group of three shadowmask holes at a time.

Are there any disadvantages with the shadowmask tube?

Much of the beam current instead of passing to the screen of the tube to activate the colour phosphor dots is intercepted by the shadowmask. This means that the efficiency of the tube in terms of light output for a given beam current is much less than that of a monochrome tube. In fact the holes in the shadowmask account for only some 15 per cent of the total area of the shadowmask so that only this percentage of the beam current lands on the screen. For this reason it is not recommended that the tube is viewed in strong ambient light. This inefficiency also accounts for the high e.h.t. used with the shadowmask tube—25 kV is the usual potential—and the much greater beam currents compared with monochrome tubes.

Is there any colour picture tube which is more efficient in this respect?

Yes, there is. The more recent Trinitron tube, for example, does not use a shadowmask; instead there is a metal aperture grille and since this has one vertical slot for each phosphor "triad" an electron transparency about 30 per cent greater than the shadowmask is achieved. The Trinitron and the "precision in-line" (PI) tube, which has an electron transparency similar to that of the shadowmask, are looked at (along with the new Mullard/Philips in-line gun tube) later in this section. But let us keep with the shadowmask tube for a while longer.

Why is it necessary to stabilise the e.h.t.?

Since variations in e.h.t. result in alterations in the size of the picture, poor regulation will affect the convergence of the three beams with the shadowmask, also the focusing, which is more critical in a colour c.r.t.

How is stabilisation achieved?

As in monochrome receivers, it is usual to derive the e.h.t. voltage from rectified stepped-up line flyback pulses. In early receivers stabilisation of the e.h.t. voltage was provided by a special shunt stabilising triode connected across the e.h.t. feed to the final anode of the picture tube. Modern receivers, however, no longer use this kind of stabilisation. Some more recent receivers, particularly those employing a transistor or transistors in the line output stage, incorporate an e.h.t. driver circuit (which itself is driven by the line timebase), the output transistor of which receives more or less power as governed by the beam current demands of the picture tube. The circuit supplying the power to this transistor is within a "control loop" whose input stages detect changes in beam current in terms of voltage across a low value series resistor at the "earthy" end of the e.h.t. supply circuit. Thus as the demand for beam current increases (bright picture) so more power is turned on by the supply part of the circuit, and as the picture goes less bright (on average) so the input power is reduced. The overall effect is tantamount to an e.h.t. supply circuit of relatively low source impedance.

Other schemes are also used, including stabilisation of the main d.c. supply to the line timebase and other primary circuits.

What are the other schemes?

The latest idea is based on a tripler circuit whereby the e.h.t. voltage of about 25kV is obtained from a relatively low amplitude line flyback pulse, the pulses at the appropriate tapping on the line output transformer being in the order of 8·4kV. This means that the resistance of the e.h.t. overwind is lower than it would be on a winding delivering a full 25kV pulse amplitude, which itself enhances e.h.t. regulation.

Further, the e.h.t. overwind is tuned to the fifth harmonic of the line frequency which flattens the tops of the pulses, thereby holding the required value of e.h.t. energy for a longer period, which is another factor of e.h.t. regulation enhancement. Moreover, improved regulation at low beam currents is sometimes attained by the use of a diode and resistor/ capacitor network at the input to the tripler, the function of the circuit being to suppress negative-going line waveform overshoot at the start of a scanning stroke by the diode conducting. Since the line waveform overshoot is not then processed by the tripler improved e.h.t. regulation at low beam currents results.

A typical e.h.t. tripler circuit is given in Fig. 19(a), which also shows the feed to the picture tube focus electrodes, the requirement here being for about 5kV. Sometimes a separate e.h.t. rectifier and separate tapping on the line output transformer are used for the focus potential, but better tracking of

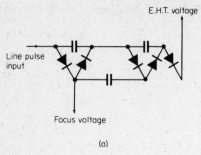

Line pulse input

E.H.T. voltage

Focus voltage

(a)

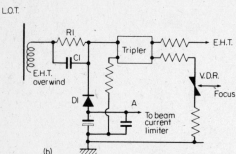
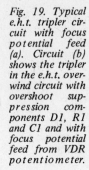

L.O.T.

E.H.T. overwind

R1

Cl

Tripler

E.H.T.

V.D.R.

Focus

DI

A

To beam current limiter

(b)

Fig. 19. Typical e.h.t. tripler circuit with focus potential feed (a). Circuit (b) shows the tripler in the e.h.t. overwind circuit with overshoot suppression components D1, R1 and C1 and with focus potential feed from VDR potentiometer.

the e.h.t. voltage and the focus voltage is obtained when the focus potential is derived from the e.h.t. supply circuit, which is another factor in the regulation equation. Fig. 19(b) shows the overshoot suppressing diode D1, along with the required resistor and capacitor R1 and R2. This circuit shows that the focus potential is obtained from a voltage-dependent resistor, arranged as a preset potentiometer. Beam current limiting is dealt with later.

Is there any danger of X-ray emission from colour television e.h.t. supply circuits?

Soft X-rays are emitted by thermionic devices in which high velocity electrons operate. The early stabiliser triode was particularly vulnerable in this respect, so to avoid the emission reaching viewers or service technicians this stabiliser was placed in a special protective screening.

Under certain conditions X-rays can also be radiated by a thermionic e.h.t. rectifier, which is the reason, again, why protective screening is provided for these — also to avoid line timebase radiation and hence undue broadcast interference. A certain amount of X-ray radiation also occurs in the picture tube, but is effectively absorbed by the thick glass and the screening surrounding the flare.

Under normal conditions, therefore, a colour television receiver does not constitute X-ray emission danger.

How is the shadowmask tube controlled?

Fig. 20(a) shows by a block diagram the arrangement generally used to control a shadowmask tube when colour-difference drive is used. A stage of video-frequency (v.f.) amplification following the video detector feeds the luminance (Y) channel, via a delay line (not to be confused with the PAL delay line), the decoder and sync separator stage. A small delay line—with a delay around $0.6\mu s$—is required to keep the luminance and "colouring" signals in step so they arrive at the display tube at exactly the same time. If this did not happen the luminance and colour components of the display would be displaced slightly horizontally.

The luminance delay is required because the colouring signals operate in circuits of lesser bandwidth than the Y signals, and as the bandwidth of a circuit is reduced (such as by the bandpass characteristic of the chroma channel and the chroma response characteristic of the i.f. channel) so the time that it takes a signal to pass through the circuit is increased, the relationship being the reciprocal of the bandwidth. The luminance channel also incorporates a notch filter tuned to the subcarrier frequency whose purpose is to reduce the level of the chroma signal in this channel. With colour-difference drive, the luminance signal is fed simultaneously to the three cathodes of the shadowmask tube, while the three colour-difference signals are fed separately to the grids of the appropriate colour guns. The decoder, of course, provides from the

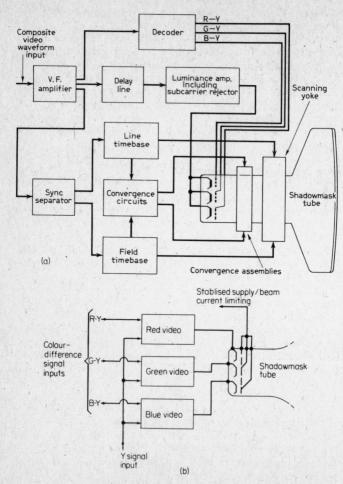

Fig. 20 Block diagram (a) of the stages required to control a three-gun shadowmask tube with colour-difference drive. (b) shows the arrangement when RGB tube drive is used.

received chroma signals separate R—Y, G—Y and B—Y colour-difference signals. The timebases yield the scanning power required for the deflection coils (scanning yoke) as in mono-chrome receivers, and in addition provide the correction currents required for the convergence assemblies mounted behind the scanning yoke on the tube neck, thereby ensuring accurate convergence of the three beams as they are deflected over the entire area of the shadowmask.

With RGB drive, commonly used in modern receivers, the three grids are connected to a "datum" potential while the three cathodes are fed separately with appropriate primary colour signals which are obtained from prior matrixing of the colour-difference signals and the Y signal, as shown in Fig. 20(b). The remaining parts of the circuit are essentially as shown at (a). The red, green and blue video stages shown at (b) consist of separate primary colour matrices and post amplifica-tion to boost the signals separately for cathode drive.

The three grids are connected together and returned to a fairly stable potential so that each gun is correctly biased between grid and cathode. Beam current limiting may also be tied in with the tube grid circuits.

What is meant by beam current limiting and how does it work?

Beam current limiting automatically ensures that the tube is not called upon to produce an excessive or dangerous amount of beam current during peak white picture information or in the event of the brightness control being too far advanced. The circuit works by effectively "sampling" the beam current and back-biasing the picture tube (all three guns) when the current rises above a pre-determined value.

From Fig. 19, for example, a "sample" of the e.h.t. is fed to a transistor circuit and hence to the tube grids, as shown in Fig. 21(a). It will be seen that the grids are connected to a positive supply through R1 and R2 which makes D1 conduct and clamp the grids to chassis. The clamping current is about 0·7mA.

TR1 is the beam current control stage which receives at its base a potential from R3/4 proportional to the beam current.

When the beam current is normal TR1 does not conduct. However, as the beam current increases TR1 switches on and when the current exceeds about 1mA (the maximum safe value) current from TR1 collector flows through R2 and D1 in opposition to the clamping bias, which switches off D1, which causes the grids to go less positive, an action which back-biases the guns and reduces the brightness so that the beam current is restored to the safe value. There are other circuits which are similar, though differing in detail (also see Fig. 23(a). TR2 is concerned with field blanking, looked at later.

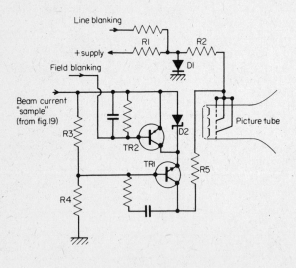

(a)

Does a colour receiver differ from a black-and-white one in the stages preceding the vision detector?

Since the chroma signals are transmitted within the normal channel bandwidth there is little need for changes in the circuits preceding the vision detector. As the chroma signal band-width is appreciably less than that of the luminance signal, however, local oscillator stability is of increased importance, and for this reason some colour receivers incorporate automatic frequency control circuits, the a.f.c. potential being used to control a variable capacitance diode in the oscillator circuit.

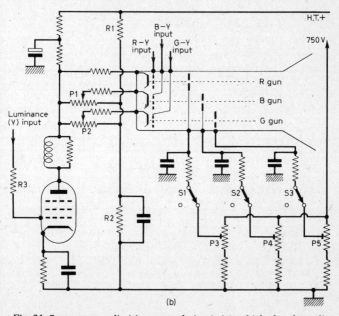

Fig. 21. Beam current limiting control circuit (a), which also shows line and field blanking feeds (see also Fig. 22). Circuit (b) is a luminance output stage of a receiver using colour-difference drive. When RGB drive is employed the luminance stage feeds Y signal to the primary colour matrices (see Fig. 20(b)).

41

Another variation found in some chassis is to tap off the chroma signal prior to the final vision i.f. stage and feed it via a separate chroma i.f. stage to a separate chroma detector stage.

How does the shadowmask tube reproduce black-and-white only pictures?

As we have seen the luminance signal provides the basic information on the brightness of the scene being scanned by the camera, being derived from the three primary colours in the correct proportions to give a signal representing black (no illumination) through the various degrees of grey to peak white. If this signal, which is exactly the same as the video signal in black-and-white television, is applied to each gun in the shadowmask tube simultaneously, the result is a black-and-white picture. What happens is that the three colour phosphors on the screen are each simultaneously activated by the beams, the light outputs adding together to give monochrome reproduction. In the arrangement shown in Fig. 21 the luminance signal is applied to the three cathodes of the shadowmask tube simultaneously. The colour circuits must be cut off during monochrome reception to prevent spurious colour patterning due to interference picked up and passed to the decoder, and for this reason a colour killer stage is included in the decoder.

Are there any snags with monochrome reproduction?

Two factors introduce complications in practice. First the efficiency of the three phosphors in translating beam current into light output differs somewhat. Because of this it is necessary, as can be seen from the valve type luminance output stage in Fig. 21(b) (now found mostly in older receivers), to provide preset controls in the feeds to the cathodes of the shadowmask tube. In the arrangement shown the R gun is fed direct but in the feed to the G and B guns are included presets P1 and P2 which apply a bias derived from the potential divider R1, R2 across the h.t. supply. In some receivers a preset control is incorporated in each feed. Since the red phosphor is the least efficient, the other signals are adjusted in relation to this to give a colourless picture *at peak white.*

The second difficulty is that it is not possible in practice to produce guns with exactly identical characteristics. This mainly affects the output at low signal levels where curvature of the Ia/Vg characteristic of the guns sets in. This problem is overcome by incorporating presets—P3–P5 in Fig. 21—to enable the first anode voltage of each gun to be adjusted independently. The adjustment is made so that each beam is just cut off with the luminance *at black level.*

The procedure of setting these controls to obtain from a shadowmask tube a monochrome picture with no colouration is called grey-scale tracking. The adjustments should be made in near or total darkness with the colour "killed" (e.g. remove the input to the decoder). The highlight adjustments should give Illuminant D white.

Switches S1-S3 enable the guns to be switched off independently, a requirement for certain tube setting-up adjustments. Receivers with RGB drive have similar controls for the first anodes, though owing to the change in nature of the Y channel there is some difference in highlight drive adjustment.

How is flyback blanking achieved?

Flyback blanking may be achieved as in monochrome practice by applying flyback pulses to a tube electrode to cut the beams off during the line and field flyback periods. An alternative approach used in a number of chassis, however, is to carry out flyback blanking in the cathode circuit of the luminance output stage. Fig. 22 shows the circuit. The transistor in the cathode circuit of the luminance output valve is normally held fully conducting by the bias applied to its base via R1. It thus has no effect under these conditions on the operation of the stage. However during the flyback periods negative field and line flyback pulses are applied to its base and cut it off. Thus during these periods the cathode of the luminance output stage is virtually open-circuit and the stage cut off. D1 protects the emitter junction of the transistor.

The scheme shown in Fig. 21(a), or one like it, is often found in receivers with RGB drive. Here negative-going line flyback pulses are fed directly to the grids, these taking clamping diode D1 out of conduction during the pulse period, thereby allowing the grids to be swung heavily negative. For field blanking positive-going pulses are used and applied to TR2 base. During the pulse period, therefore, TR2 conducts and shorts zener D2. This causes TR1 current to rise sharply, so that negative-going pulses are communicated to the grids through R5.

How is brightness control achieved?

Since in the type of arrangement depicted in Fig. 20(a) all the shadowmask tube control grids and cathodes are used for beam modulation purposes it is not possible to adjust the brightness by setting the d.c. bias applied to the grid with respect to the cathode or vice versa as is the practice with monochrome receivers. Thus the brightness control, still necessary to set the overall picture brightness, must be placed farther back in the circuit. The technique with colour-difference drive receivers is to incorporate the brightness control in the control grid (or base) circuit of the luminance output stage, operating it in conjunction with the d.c. restorer used at this point following a.c. coupling to the grid (or base). A refinement found in many receivers here is a beam current limiting circuit to reduce the tube beam current should it for any reason, for example, incorrect setting of the contrast and brightness controls, exceed the range of control of the e.h.t. stabilisation systems described earlier. A simple example in shown in Fig. 23(a) and this type of circuit is used in a number of models. The cathode side of diode D1, which is normally cut off, is taken to the grid circuit of the shunt stabiliser triode in the e.h.t. circuit. As we have seen, at high beam currents the shunt stabiliser triode grid is driven negative; when it is sufficiently negative, D1 conducts as it is then forward biased. As a result the increased current through R1 and the higher voltage across it reduces the voltage across the brightness control and consequently the grid voltage of the luminance output stage and hence the drive applied to the tube. D2 is the d.c. restorer diode. A number of other techniques are now used, particularly in receivers with RGB drive (see Figs. 19(b) and 21(a)). A common technique is to sample the beam current at the cathode of the line output valve. Control can be applied in a number of ways, one less common arrangement being via the a.g.c. line.

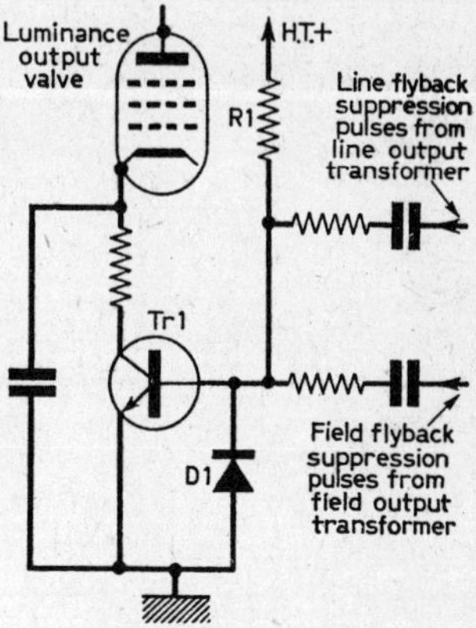

Fig. 22 (right). Line and field flyback suppression in the cathode circuit of the luminance output stage.

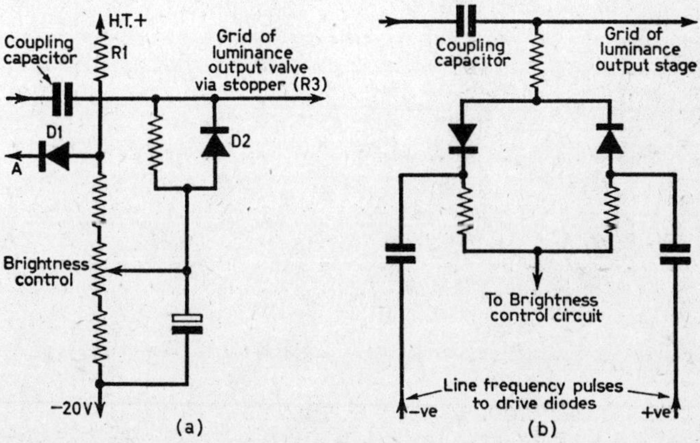

Fig. 23 (below). (a) Features of the luminance output stage grid circuit: brightness control, d.c. restoration following a.c. coupling, and a beam limiter arrangement to prevent excessive beam current. (b) Double-diode driven clamp circuit, used here to provide black-level clamping.

45

A further refinement in one chassis is the use of a two-diode driven clamp in the luminance output stage grid circuit instead of a simple d.c. restorer. The arrangement is shown in Fig. 23(b). The diodes are driven by positive- and negative-going line frequency pulses and the circuit provides improved black-level clamping.

An alternative approach to brightness control that has been used is to incorporate a potentiometer for this purpose in the screen grid circuit of the luminance output valve.

How about receivers using RGB drive?

The control of brightness in receivers using RGB drive is often related to the clamping diodes on each of the primary colour feeds to the picture tube cathodes. A variety of schemes are adopted, but they are all mostly based on the principle that the brightness control regulates the clamping level in such a way as to change the biasing of the guns. More information about this is given under a later question.

How is contrast control effected?

In some receivers contrast control is effected as in monochrome receivers by setting the level at which the a.g.c. action commences. An alternative approach is based on a potentiometer between the Y input and the Y delay line such that by adjustment to the potentiometer the level of the Y signal is regulated. There are circuits, too, which gang the contrast control to a "colour compensation" control, so that as the control is adjusted the colour drive varies in accordance to the degree of contrast used. Moreover, in one receiver, at least (ITT CVC5), beam current limiting is geared to the luminance drive (contrast control circuit) in such a way that when the potential across a low value resistor in series with the line output stage rises (indicating a rise in tube beam current), a controlling effect automatically retards the Y drive, thereby reducing the demand for excessive beam current.

Does the application of a.g.c. present any problems?

The mean-level system used on nearly all 405-line receivers

is not suitable for colour reception on 625 lines. Its well-known defect, that the control potential it provides varies with picture content as well as signal strength, means on black-and-white that the black level is not maintained and on colour in addition that the intensity of the colour reproduction would vary. For this reason an a.g.c. system that samples the waveform at some definite level is preferable. With the negative modulation used on the 625-line system the tips of the sync pulses represent maximum signal amplitude and provide a convenient point at which to sample the signal. The sync tip a.g.c. system is, in fact, used on all current colour television chassis.

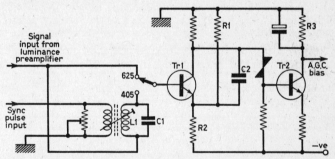

Fig. 24. Basic sync tip a.g.c. circuit with sync amplifier stage (Tr2) and sync ringing circuit (L1, C2) to provide gated a.g.c. on 405 lines

How does the sync tip a.g.c. system work?

Either a diode or transistor may be used as a peak detector to sample and rectify the signal at the sync tip level to provide the a.g.c. potential. A widely employed circuit using a transistor as peak detector is shown in Fig. 24. The composite luminance signal is fed to the base of the transistor, its emitter being biased by resistors R1 and R2 so that it only conducts on the tips of the sync pulses. As a result it provides an output which is a sample of the waveform at a definite level, that of the sync tip, and which is independent of picture content. The output pulses are smoothed by C2 and used to control the a.g.c. amplifier transistor Tr2. As with hybrid black-and-white receivers forward a.g.c. is used (in which the a.g.c. potential increases the conduction of the controlled stage on increase in signal level, thereby reducing the collector voltage because of the inclusion of a series resistor in the collector lead) and the a.g.c. amplifier is bottomed (fully conducting) on weak signals. With increase in signal strength the output from the rectifier reduces the current in Tr2 and the voltage across R3, its load resistor, falls, i.e. is positive going. This positive-going potential, in the case shown, is used to bias the base of an npn controlled stage, thereby increasing its conduction (npn transistors require a positive-going drive at their base for increased conduction). Most colour receivers use silicon npn transistors for the r.f. and i.f. stages. Where germanium pnp transistors are used in the controlled stage a pnp a.g.c. amplifier transistor is employed and a negative-going bias is applied to reduce stage gain in the controlled stage on increase in signal strength.

What happens on 405 lines?

Clearly the sync tip a.g.c. system cannot be used on 405 lines since the sync tips then represent nil modulation and maximum modulation is peak white. A few colour sets on 405 reverted to mean-level operation. In a number of chassis, however, a gated system is used in conjunction with the 625-line sync tip system. Reverting to Fig. 24, on 405 lines the input to the a.g.c. rectifier is taken via the tuned circuit L1, C1. To this is coupled a winding fed with pulses from the sync separator. These sync pulses provide the gating action, the circuit being arranged so that the rectifier conducts only on the first positive overswing following the sync pulse, i.e. at the black level of the picture (the back porch period) on the 405-line system. This technique is called the sync pulse ringing circuit.

How is colour reproduction obtained?

We have now seen how the shadowmask tube, controlled as shown in Fig. 20, reproduces black-and-white pictures when the three beams in the tube are modulated by applying the luminance signal simultaneously to the three cathodes. To

48

introduce colour into the picture, all we have to do is to vary the modulation of the beams further; for example to obtain a red picture we need to increase the magnitude of the red beam so that the red phosphor dots on the screen glow more brightly to give an overall red display, while reducing the output from the other beams to maintain the correct brightness level.

This is where the colour-difference signals come in. These, in the arrangement shown in Fig. 20, are applied to the three grids. Let us as before take the case of the red gun, and suppose that the scene being televised contains red light so that a red signal is provided by the red-sensitive camera tube in the television camera. Now some of the red signal is already present at the cathode in the luminance signal since, as we saw in Section 1, this contains 30 per cent of the output from the red-sensitive tube in the camera. So what we need to obtain on the screen the true "redness" of the scene being scanned is the difference between the red in the luminance (Y) signal and the actual red signal output from the red camera tube, in short we need an R−Y signal, or 70 per cent of the red signal, to add to what we already have. This signal, then, obtained from the decoder which derives it from the transmitted chroma signal, is applied to the grid of the red gun in such a manner as to add to the Y signal at the cathode so that on the screen we obtain the actual red of the original scene. At the grids of the other two guns the colour-difference signals fall to reduce their output—or to cut them off completely—so that the brightness remains in accord with the Y signal.

The same principle applies to each of the other primary colours, so that the three grids, on a colour transmission, will be fed with the R−Y, G−Y and B−Y information they need to give a full colour picture, the proportions of the signals constantly changing to give the varying shades present in the scene being televised at the correct brightness level.

It is important to note that the bandwidths of the luminance and colour-difference signals differ markedly—about 5·5 MHz in the former case and roughly ±1 MHz in the latter. This means that the *detail* of the picture is provided by the luminance signal on both monochrome and colour, the colour-

49

difference signals bringing out the colour but not providing fine detail as to changes in colour. As we saw in Section 1, this is acceptable because the human eye is not sensitive to fine detail in respect of colour.

Why is it necessary to clamp the three grids?

It is necessary to maintain the correct d.c. conditions at the grids as well as at the cathodes. What this means in practice is that the bias on the grids with respect to the cathodes must be maintained such that the bias at each gun just cuts off the beams at black level. If this condition is not maintained the contrast range of the picture will not be correct, that is grey-scale tracking will be affected, and colour reproduction will also be incorrect. The luminance signal is d.c. coupled to the cathodes of the shadowmask tube, but a.c. coupling is generally used to the grid (or base) of the luminance output stage so that d.c. restoration is, as we have seen, incorporated at this point. An additional precaution taken in one chassis is to include a zener diode in the cathode circuit of the luminance output stage.

A.C. coupling is generally used at some point in the colour-difference signal path. For example in the widely used arrangement shown in Fig. 25 the colour-difference output stage comprises a pentode (triodes are also used for this stage) a.c. coupled to the shadowmask tube grid (three such stages in all, one for each colour-difference signal, are used). This means that the grids must be clamped to the black level and in the circuit shown a triode clamp is used. This is fed with positive-going pulses from the line output stage, and operates in conjunction with the d.c. bias resistor R1 to provide grid clamping. The line frequency pulses open the clamp when the signal is at black level, i.e. during the back porch period, to restore the voltage across R1 to the level set by the potential divider R2, R3.

How do RGB drive clamping and brightness control operate?

Fig. 26 shows the Y circuit and the red channel of a receiver with RGB drive, along with the appropriate clamping

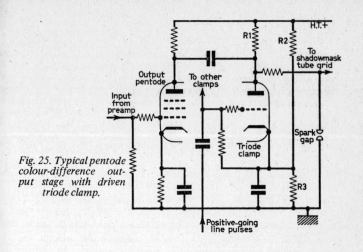

Fig. 25. Typical pentode colour-difference output stage with driven triode clamp.

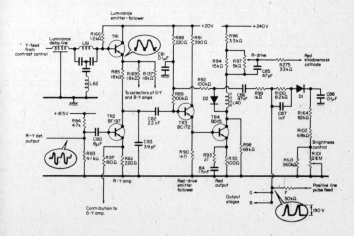

Fig. 26. Circuit of red channel in receiver using RGB drive, showing brightness control and clamping circuit. See text for description.

51

circuit and the brightness control. Here Y signal from TR1 emitter is added to R—Y signal from TR2 collector, the addition then being applied to TR3 base. Since TR3 is an emitter-follower, red signal appears at the emitter and this is directly coupled to the base of the red output transistor TR4, whose collector drives the red shadowmask cathode.

Owing to the a.c. coupling by C80 and C82 clamping is necessary to stabilise the d.c. per line at the red cathode (also at the green and blue cathodes, which are dealt with similarly), and this is achieved by the "brightness control" diode D1 and the clamping diode D2 (diodes similar to D2 are also used in the green and blue output circuits). At line flyback a positive-going pulse is fed to D1 anode such that the diode conducts when the charge on C88 is less than the pulse peak amplitude. Since the charge on C88 is regulated by the brightness control, it follows that the pulse is clipped to a value determined by the setting of the brightness control. The clipped pulses are timed so they occur during the blanking periods of picture signal when they cause conduction of the clamping diode, thereby clamping D2 cathode to TR4 collector.

This means that the potential at D2/R92 junction reflects the pulse amplitude established by the brightness control and the instantaneous signal voltage at TR4 collector. As this potential is in communication with TR3 base a d.c. clamping action occurs such that the signal rides on the d.c. which is biasing the gun. Thus as the brightness control is adjusted so also is the biasing and hence the blanking level brightness.

What sort of colour-difference drive voltages are required?

With colour-difference drive, where the guns perform primary colour matrixing, the beam current yielded by each gun is determined by the net instantaneous voltage between cathode and grid. Since the cathodes are driven by Y signal and each grid by the appropriate colour-difference signal it is obvious that the amount of colour-difference drive required at each grid must depend on the amount of the appropriate colour-difference signal there is in the Y signal. To produce white light approximately equal intensities of red, green and blue light are required. Assuming equal efficiencies of the three phosphors, this would call for equal beam currents.

In practice, there are complications because first of all the efficiences of the three phosphors are different and secondly a greater change in beam current occurs when a signal of given amplitude drives the cathode than when it drives the grid; the sensitivity of cathode drive is about 30 per cent greater than grid drive. However, if we ignore the differences in phosphor efficiencies, we find that maximum peak-to-peak drives at the red, green and blue grids of R−Y 182V, G−Y 106V and B−Y 232V are required from the respective colour-difference amplifiers. The G−Y voltage swing is least because there is more green component in the Y signal than there is red or blue. The voltages are approximately in proportion to the amount of corresponding colour in the Y signal.

In the final tailoring the different efficiencies of the three phosphors are also taken into account; in general, the blue phosphor is the most efficient and the red phosphor the least efficient.

What about RGB drive?

Since the matrixing is performed prior to the picture tube the Y "cancelling" effect with this sort of drive does not occur as with colour-difference drive. Cathode drive is commonly employed owing to its greater sensitivity than grid drive. This results because as the cathode of a gun is driven less positive (to increase beam current), the effective voltage at the first anode relative to cathode goes more positive, thereby further increasing beam current. This, of course, does not occur with grid drive because then the cathode voltage is "fixed". As already mentioned, cathode drive provides a sensitivity increase of about 30 per cent over grid drive. Small wonder, then, that it is favoured by designers since it reduces the drive demands imposed on the primary colour amplifiers.

A colour tube may require about −100V at each cathode for peak white output, but when the differences in phosphor efficiencies are taken into account for correct white the red gun may require, say, −100V, the green gun −95V and the blue gun −90V.

What is the effect of the saturation control?

All colour receivers are fitted with a saturation control (often called the colour control) to enable the overall colour strength of the picture to be adjusted by the viewer. The control adjusts the gain of the chrominance amplifier stages in the decoder. The colour and contrast controls are sometimes linked so that they track together.

What is meant by a tint control?

Several chassis incorporate a tint control which enables the colour balance to be adjusted, increasing the red output whilst reducing the blue output or vice versa to enable a warmer or colder overall picture to be obtained.

What causes incorrect colour reproduction on the screen?

Apart from faults in the electronic circuits there are several shadowmask tube setting up adjustments which, if incorrect, will mean that correct colour reproduction is not achieved. These are the purity and convergence adjustments. There is also the need for degaussing. All these adjustments are to ensure that the three tube beams activate only their respective phosphors on the screen. Clearly, for example, if the direction of the red beam is incorrect it may strike, say, part of the green dots, making the colour reproduction incorrect, and since the strength of the three beams differs to take into account the different efficiencies of the three phosphors, such an error can have a marked effect on colour reproduction. The position of the purity and convergence assemblies on the neck of the shadowmask tube is shown in Fig. 27.

What is meant by degaussing?

Degaussing is the removal of residual magnetism in the receiver, since such magnetism would interfere with the correct scanning of the three beams. A degaussing coil may be used to demagnetise the set on installation and this must be

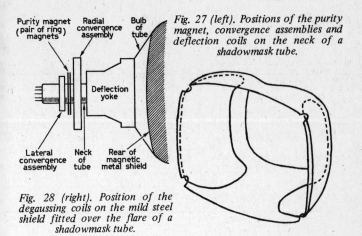

Purity magnet (pair of ring magnets) · Radial convergence assembly · Bulb of tube · Deflection yoke · Lateral convergence assembly · Neck of tube · Rear of magnetic metal shield

Fig. 27 (left). Positions of the purity magnet, convergence assemblies and deflection coils on the neck of a shadowmask tube.

Fig. 28 (right). Position of the degaussing coils on the mild steel shield fitted over the flare of a shadowmask tube.

done with the set in the position in which it is to remain. In addition all sets incorporate automatic degaussing coils wound, as shown in Fig. 28, on the mild steel shield fitted round the shadowmask tube and brought into operation each time the set is switched on (in addition in some models degaussing is also carried out when system switching). These coils demagnetise the main metal parts of the shadowmask tube—the shadowmask, the magnetic shield and the mounting band. The basic principle of automatic degaussing is shown in Fig. 29(a). The degaussing coils are connected across the mains input to the set in series with a thermistor having a positive temperature coefficient. Thus when the set is switched on an a.c. field is established around the coils and is of sufficient magnitude to overcome the residual magnatism in the metal parts of the tube. This field is gradually reduced to nearly zero by the action of the p.t.c. thermistor, leaving the atomic structure of the metalwork "jumbled", i.e. no longer magnetically orientated.

On switching on a substantial a.c. flows through the coils, the p.t.c. thermistor then having a very low resistance. As the

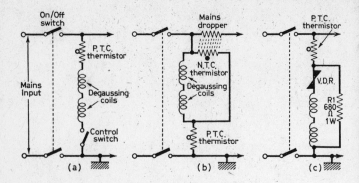

Fig. 29. Automatic degaussing. (a) Principle, (b) and (c) two widely used fully automatic arrangements

p.t.c. thermistor warms up its resistance increases, reducing the current flowing through the coils to a very low value. The control switch is then opened to take the degaussing network out of circuit. Rather, however, than use a separate switch electronic means are used to render the degaussing coils inoperative after the tube has been demagnetised. Of a number of possibilities two techniques in use are shown in Fig. 29(b) and (c). In (b) a thermistor with a negative temperature coefficient shunts the coils. This thermistor is mounted close to the mains dropper resistor so that it heats up rapidly, thus quickly falling to a low resistance value. After degaussing the situation is that almost all the voltage develops across the p.t.c. thermistor, the n.t.c. thermistor shorting out the coils. In (c) a voltage-dependent resistor is connected in series with the coils, the v.d.r. and coils being shunted by a resistor (R1). On switching the set on the demagnetising current flows through the coils and v.d.r., but as the p.t.c. thermistor warms up and the voltage across it increases so also does the resistance of the v.d.r. As a result the coils and v.d.r. form a high-resistance path and are shorted out by the shunt resistor R1.

56

What is the purity adjustment?

To maintain the purity of colour reproduction the three beams must activate only their corresponding phosphors. This means that their angle of approach must be correct so that after passing through the shadowmask they fall on the centres of the corresponding phosphor dots on the screen and do not overlap even slightly on to adjacent phosphor dots. Fig. 30 shows how if a beam approaches the screen at the wrong angle it can overlap phosphor dots of the wrong colour.

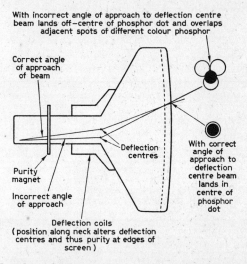

Fig. 30. How the angle of approach of the beams to the points where they are deflected affects the purity of reproduction.

External magnetic fields interfere with colour purity, and to overcome this an adjustable purity magnet is fitted on the neck of the shadowmask tube. This consists of two thin ring magnets which can be moved independently: rotating them together varies the direction of the magnetic field in the neck of the tube, whilst moving them in opposite directions varies

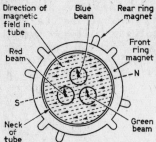

Direction of magnetic field in tube

Blue beam

Rear ring magnet

Red beam

Front ring magnet

N

S

Neck of tube

Green beam

Fig. 31. The three beams move in the direction perpendicular to the magnetic field established by the purity magnet (a pair of ring magnets). Rotating the ring magnets together alters the direction of the magnetic field, rotating them in opposite directions alters the strength of the magnetic field.

the strength of the field. The effects are illustrated in Fig. 31. The purity magnet affects colour purity towards the centre of the screen. Towards the edge of the screen, however, the position of the deflection coils is the main influence on purity so that they must be correctly positioned along the neck of the tube.

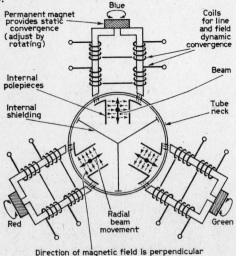

Permanent magnet provides static convergence (adjust by rotating)

Blue

Coils for line and field dynamic convergence

Internal polepieces

Beam

Internal shielding

Tube neck

Red

Radial beam movement

Green

Direction of magnetic field is perpendicular to radial beam movement

Fig. 32. Radial convergence assembly.

58

What is meant by convergence?

Not only must the angle of approach of each beam be correct, but the three beams must converge, i.e. come together, at the same point of the shadowmask. This is to ensure that the three pictures scanned out by the three beams coincide. If the beams are not in convergence colour fringes will be present on the outlines of objects in the picture because colour changes will not occur at the same point in the picture.

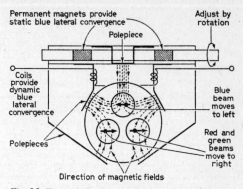

Fig. 33. Typical blue-lateral convergence assembly.

To bring the beams into convergence we need to be able to adjust their positions, independently, in two different directions. Thus two convergence assemblies are used, one producing radial movements of the three beams and the other, which affects mainly the blue beam, causing lateral beam displacement. Further complications arise because due to the geometry of a flat-faced tube the convergence correction required for correct convergence at the centre of the picture differs from that required at the edges. For this reason there are two sets of adjustments to each convergence assembly. Static convergence, as it is called, is provided by means of adjustable permanent magnets and sets the convergence correctly at the centre of the picture. Dynamic convergence to obtain correct convergence at the edges of the screen is, on the other hand, provided electromagnetically by means of coils through which appropriate correction current waveforms are fed. These currents are derived from the timebases. Convergence adjustments consist of pre-setting the permanent magnets and adjusting the amplitude and shape of the correction waveforms applied to the dynamic convergence coils. Typical radial and lateral convergence assemblies are shown in Figs. 32 and 33.

Fig. 34 illustrates how correct convergence is obtained. Radial convergence correction brings the red and green beams together and moves the blue beam on to the same lateral line; lateral convergence correction then superimposes all three beams.

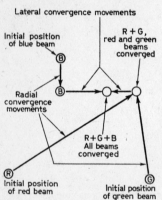

Fig. 34. How the three beams are converged. Two types of beam movement, radial and lateral, are needed. The red and green beams are converged upon each other by radial movements and the blue beam brought on to the same horizontal line in the same way. Lateral movement then converges all three beams at the same point on the shadow-mask.

How is picture centring and pin-cushion distortion correction achieved?

In black-and-white receivers picture centring and pin-cushion distortion correction are achieved by means of permanent magnets mounted around the tube. Clearly the same techniques cannot be used in a colour television receiver since such magnets would interfere with the purity and convergence of the display. Instead centring (or shift as it is called) is achieved by adjusting the d.c. component of the currents flowing in the line and field deflection coils, while pin-cushion distortion may be corrected by modulating the line deflection current with a suitable waveform (a parabola) at field frequency and modulating the field deflection current with a line frequency parabolic waveform.

What is meant by 110° picture tubes?

This refers to the angle of deflection of the electron beams required for full-area scanning of the screen. Throughout the development of picture tubes the scanning angle has been progressively increased to cater for larger screen sizes without an undue increase in tube length, and this development has now caught up with colour tubes. Of recent times the scanning angle of the shadowmask, for example, has been increased from 90° to 110°, a feature which enables a reduction in cabinet depth for a given screen size or no significant increase in depth for a larger screen size. Most 110° tubes have the degaussing shield built into them.

Are there any problems resulting from the increased scanning angle?

As the deflection angle is increased, so it becomes more difficult to control exactly the three electron beams for optimum convergence, purity and focus, but to some extent this is tied up with the thickness of the tube neck. Both thick- and thin-neck tubes have been developed, the former having a neck thickness similar to that of 90° tubes and the latter a somewhat thinner neck containing miniature and closely spaced electron guns. Because the electron beams are thus close together from the start, dynamic problems are minimised with the thinner neck tube, and a relatively simple passive dynamic convergence arrangement (as with 90° tubes) has been developed by ITT, the inventors of the thin-neck tube. The company has also developed a thyristor line timebase circuit for use with the tube.

The thick-neck tube, on the other hand, may require a more complicated convergence control system, using active circuits, and steps are generally taken to "stabilise" the focus potential which, owing to the overall tube/scanning system

61

design, may otherwise vary at line rate. However, the nature of the scanning coils allows the use of a single transistor line output stage in spite of the greater scanning angle.

How does the Trinitron tube differ from the shadowmask?

The Trinitron tube is a Japanese Sony development and is thus found in Sony colour receivers. Development and research is also attributable to such firms as Susumu, Yoshida and Akio Ohgoshi in certain aspects—all Japanese and led by the Japanese engineer Senri Miyaoka. The tube was first announced by Sony during April 1968 and is the subject of more than 100 patents.

The Trinitron differs from the shadowmask in that it employs a single electron gun and "split beam" approach. Moreover, instead of using a shadowmask it employs what is called an "aperture grille" which consists of a metal plate "etched" to the form of a grille of vertical stripes and instead of the screen being composed of a multiplicity of red, green and blue glowing phosphor dots in triad formation vertical stripes of red, green and blue glowing phosphors are used. The phosphor stripes are very close together so that at normal viewing distance the three-colour illumination at any instant merges to give the impression of the correct colour of the picture element transmitted, the function in this respect, of course, being the same as the shadowmask.

Thus each beam section excites but one colour, the arrangement being that the beams converge at and pass through the grille apertures as they are deflected vertically and horizontally in the ordinary manner. Instead of the beams emanating from the gun in delta formation, as they do from the three guns of the shadowmask tube, they appear from the single gun in line, as made clear by the elementary diagram of the system in Fig. 35, where it is shown that the beams are controlled by two electron lenses and an electron prism.

Since the aperture grille has a greater electron transparency than the shadowmask, a brighter picture than that yielded by the shadowmask tube is attainable from a given e.h.t. voltage. Moreover, the aperture grille is less influenced by extraneous

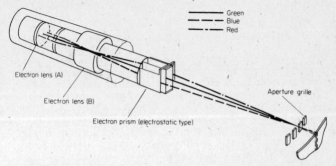

Fig. 35. Elementary impression of the Trinitron colour picture tube.

magnetic fields than the shadowmask, and because the phosphors are vertically disposed Moiré pattern interference is reduced under certain conditions.

Convergence with the Trinitron is less of a problem (with the smaller tubes, anyway) than with the shadowmask because the three beams lie on a common horizontal plane. Basically, therefore, it is necessary merely to have some means of adjusting the approach angles of the two outer beams to ensure that they converge correctly at the aperture grille with the beam emanating from the centre cathode of the gun. This adjustment is facilitated by the electrostatic electron prism. The two outer beams excite the red and blue phosphors and the middle beam the green phosphor. Thus there is an electron prism for each of the red and blue beams. The prisms work on the electrostatic principle, so that corrective beam deflection for convergence—left or right—is achieved merely by regulating the voltages applied to the prisms, this leading to relatively simple circuitry. However, as the three beams scan across the aperture grille the convergence tends to deteriorate from the nominal, so some method of dynamic convergence correction is required, as with the shadowmask tube. The correction signal is picked up from the line timebase and applied as a voltage to the prisms, the signal having the nature of that shown in Fig. 36.

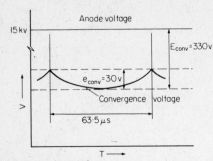

Fig. 36. Nature of the line voltage waveform required for Trinitron dynamic convergence.

It is noteworthy that the nature of the tube design gives a greater resolution of the centre green "spot" than the red and blue "spots" from the outer beams, but since the green light determines the overall definition of a colour picture this is a good thing since it ensures optimum colour resolution.

What are the basic features of the PI tube?

This tube, a U.S. development by RCA and introduced to the U.K. by Thorn under the Mazda label, is likely to appear in British receivers by the time this new edition is in print. Like the Trinitron, it also employs horizontal in-line beams, thereby confining convergence problems to horizontal shift, from a special gun assembly and also red, green and blue glowing phosphor stripes, but unlike the Trinitron the outer beams are green and blue. The idea behind this is to reduce the subjective effect of red misconvergence, as this is most noticeable, by making the middle beam excite the red phosphors.

So that the faceplate can follow the conventional spherical curvature of the shadowmask tube, the aperture does not consist of a grille of vertical slots as in the Trinitron, since this makes it possible to establish only a horizontal curvature, which is a characteristic of the Trinitron tube. Neither does it consist of a multiplicity of holes as in the shadowmask tube. Instead a compromise between holes and slots is adopted. The mask of the PI tube, therefore, is composed of vertically elongated holes, which gives it an electron transparency of similar order to the shadowmask. It has a $90°$ scanning angle.

One important feature of the tube is the deflection unit, called "precision static toroid deflection yoke". This is very accurately designed to yield an astigmatic deflection field which helps to provide automatic convergence of the three beams as they are scanned across the screen in the usual manner. For this to happen, however, the yoke needs to be very accurately positioned on the tube neck up against the flare. Thus the yoke is treated as an integral part of the tube system and is aligned very accurately at the factory and then cemented permanently in position. Thus when a tube is replaced the yoke comes complete with the replacement. It is possible, however, to remove the yoke by heating the thermoplastic cement which secures it to the tube.

Static convergence is effected by two four-pole and two six-pole ring magnets on the tube neck. There are also the usual purity magnets.

To summarise, therefore, the tube system is designed to be automatically converging, which of course is a big step towards the simplification of colour receivers. Fig. 37 shows the precision scanning yoke, which is cemented to the tube, and Fig. 38 shows a front view of the gun assembly which contains small ring magnets acting as magnetic shunts and enhancers for extra beam correction.

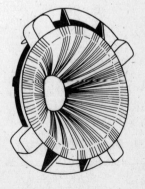

Fig. 37. The precision scanning yoke which forms a part of the new PI colour tube. This is specially engineered to provide an astigmatic scanning field for self-convergence.

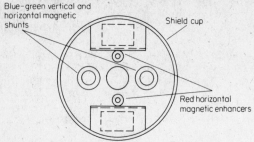

Fig. 38. Front view of the PI electron gun assembly, showing the magnetic shunts and enhancers which provide raster tailoring.

Are there other tubes with in-line guns and phosphor stripes?

Yes. Recently Mullard in conjunction with its parent Philips has produced an in-line gun tube with slots instead of circular holes in the shadowmask and a series of phosphor stripes instead of phosphor dot triads. It differs mainly from the PI tube in four ways: its scanning angle is 110° (the PI tube is 90°); it has a thicker neck; saddle-wound deflection coils are used, which are separate though accurately aligned; and all the electrodes of each gun are available separately at the base, making circuit design easier. It is also available in a whole range of sizes.

Another is the Toshiba RIS (standing for rectangular flare in-line guns and slotted shadowmask) tube. This also has a 110° scanning angle and a thick neck (36 mm). The new tubes simplify convergence and setting-up procedures, amongst other things.

3

DECODING THE CHROMA SIGNAL

What is the purpose of the decoder in a colour receiver?

The purpose of the decoder is to accept the chroma signal and process it so as to derive separate R−Y, B−Y and G−Y colour-difference signals. The colour-difference signal amplifier stages may be mounted on the decoder panel (plug-in printed panel construction is generally used for colour receivers) or on a separate "video" panel. Since G−Y matrixing (deriving the G−Y signal which is not transmitted by adding −(R−Y) and −(B−Y) signals in the correct proportions) is done in the colour-difference preamplifier or output stages, this too may be physically done on either the decoder or video printed panel.

What are the main operations carried out in the decoder?

The chroma signal has to be separated from the luminance signal, amplified, the B−Y and R−Y components separated and then demodulated by means of synchronous detectors. Since suppressed subcarrier transmission is used a reference oscillator operating at the same frequency and in phase with the subcarrier oscillator in the transmitter must be included in the decoder, the output of this being used by the decoder demodulators in the process of synchronous detection. And since the R−Y component of the signal is phase inverted on alternate lines in the PAL system a means of changing the phase of the R−Y signal back on alternate lines must be included. The PAL feature of automatically correcting phase changes in the transmitted signal by averaging successive pairs

67

of lines must be provided (unless simple optical averaging, not at present resorted to on commercial models, is used). The burst signal transmitted as a reference or sync signal for the decoder reference oscillator must be separated from the composite signal and used to control the phase and frequency of the reference oscillator; it must also be removed, or "blanked out", from the chroma signal path to prevent it being demodulated and interfering with the output stage black-level clamping. And since the phase of the burst signal changes on alternate lines because of the alternate line R−Y signal phase inversion, this "swinging burst" feature, as it is called, is used to generate an ident signal to ensure that the alternate line R−Y signal inversion in the decoder is in step with that at the transmitter. Ident is thus short for R−Y alternate line phase reversal identification. Matrixing to recreate the G−Y colour-difference signal must be undertaken. And finally a means must be provided to make the decoder inoperative on monochrome reception to avoid colour patterning on black-and-white displays. For this reason a colour killer circuit is included. A.C.C. is nearly always incorporated.

Is the design of decoders standardised at all?

Whilst the same basic series of operations has to be carried out, and many decoders follow a similar pattern, there are nevertheless considerable differences in decoder techniques, as we shall see. For example R−Y signal inversion may be done by inverting the received and separated R−Y signal on alternate lines before feeding it to the R−Y synchronous detector, or the signal from the reference oscillator to the R−Y synchronous detector may be inverted on alternate lines. Then while in most decoders a bistable circuit is used to control the alternate line R−Y inversion, an alternative approach is to amplify the ident signal, square it and use it directly to control the alternate line R−Y inversion. There are variations in the gating, blanking and clamping techniques used, and a number of different synchronous demodulator and burst detector circuits are in use.

What are the basic signal paths in the decoder?

A PAL decoder reduced to its basic essentials is shown in block schematic form in Fig. 39. The composite video waveform, which may be taken off from a video frequency stage in the luminance channel (as in Fig. 20), the video detector, or from a separate chroma detector fed from a separate chroma i.f. stage, is applied to a chroma amplifier stage which separates and amplifies the chrominance part of the applied signal and uses this to drive and delay line and B—Y, R—Y signal separation network. The separated B—Y and R—Y outputs are detected by separate synchronous demodulators, and the G—Y signal obtained from a matrix circuit driven by these two outputs. The three colour-difference outputs require further amplification before application to the shadowmask tube.

Two other inputs are applied to the chroma amplifier. First a bias input from the colour killer stage. This switches the chroma amplifier on when a colour transmission is being received, and off during a black-and-white transmission to ensure that spurious signals entering the decoder do not produce colour patterning on monochrome displays. And secondly a pulse input to bias the stage off during the burst signal to remove this from the chroma signal.

The input signal to the decoder is also applied to a burst gate/amplifier stage. This is gated on by the line pulse input so that only the burst signal, amplified, appears at its output. This is fed to a phase detector circuit which compares the phase and frequency of the burst signal with that of the decoder reference oscillator. The phase detector provides a control bias, in the same manner as in flywheel line sync circuits, and this is used to adjust a variable reactance in the decoder reference oscillator circuit. Since the subcarrier frequency must be precisely controlled for correct colour reproduction, a crystal oscillator is used as the reference oscillator in the decoder, automatically controlled via the phase detector circuit as just indicated.

The reference oscillator output is used to control the synchronous demodulators, switching them on at appropriate

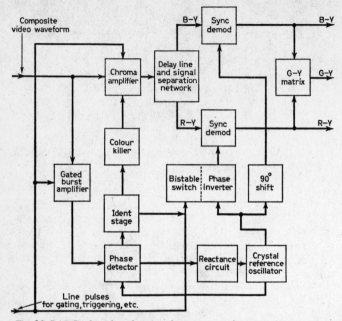

Fig. 39. PAL-D decoder, showing the processes that must be carried out to obtain the three colour-difference signals from the transmitted chroma signal and the signal paths in the decoder.

intervals to detect the B−Y and R−Y signals. Since at the transmitter these signals modulate the subcarrier in quadrature, i.e. with a phase difference of 90° between them, as we saw in Section 1, a 90° phase-shift network is included in the feed to one of the synchronous demodulators so that the reference oscillator signals to the synchronous demodulators are 90° apart in phase as at the transmitter.

In the example shown, the more usual technique of inverting the phase of the R−Y reference oscillator feed to the R−Y synchronous demodulator on alternate lines is used. This is done by means of a bistable switch circuit which switches the

70

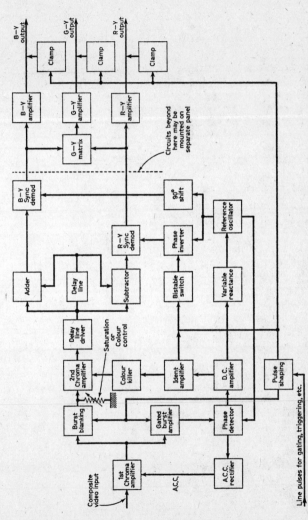

Fig. 40. Block diagram of a practical PAL-D decoder. This type of arrangement is very widely used.

71

phase of a phase inverter circuit on alternate lines. The bistable switch is triggered by line frequency pulses.

Since, as we have seen, the burst signal varies in phase from line to line (±45° about the reference point), a signal at half line frequency (7·8 kHz) appears in the output of the phase detector circuit. This is the ident signal, and is used to excite a high-Q resonant circuit tuned to this frequency. The resultant 7·8 kHz ident signal is used for two purposes. First to control the triggering of the bistable switch so that it corresponds with that at the transmitter. And secondly the ident signal is rectified, smoothed and used as the colour killer bias potential. Since the ident signal is only present on a colour transmission, the presence or absence of the ident signal is a convenient basis for colour killer operation.

How does a practical decoder differ from the scheme just outlined?

Fig. 40 shows in block form a widely used decoder arrangement. Two chroma amplifier stages are generally used, plus in addition a delay line driver stage. Burst blanking is, as we have seen, carried out in the chroma channel, where the saturation (or "colour") control to provide user control over the strength of the colour signal is also generally situated. In addition one of the chroma stages, generally the first, has a.c.c. (automatic chrominance control) applied to it. A common a.c.c. technique is to rectify and smooth the burst signal appearing in the phase detector circuit, using this as the a.c.c. potential. Two burst amplifier stages are generally used, either the first or the second being gated to remove the chroma signal information and allow through only the burst signal.

The technique used in the delay line and B−Y, R−Y signal separation network is outlined in Fig. 40. The output from the delay line driver stage is fed directly to adder and subtractor networks and is also fed to these networks via the line-period (approximately 64μs) delay line. By adding and subtracting the two sets of chroma signals, one obtained direct from the driver and the other via the delay line, separate B−Y and R−Y outputs are obtained. These outputs are the average taken over

72

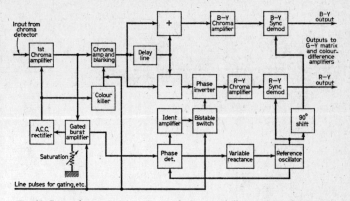

Fig. 41. Some alternative decoding techniques are illustrated in this PAL-D decoder block diagram. The main change is the inclusion of the alternate alternate line R−Y switching in the R−Y signal path.

two successive lines and as a result of the phase inversion of the R−Y component of the composite signal on alternate lines spurious phase shifts cancel out in the output obtained from these networks.

Line frequency pulses obtained from the line output stage may be used for four different purposes in the decoder: for burst blanking, burst gating, to trigger the bistable circuit, and to clamp the colour-difference signal outputs. The pulses used for burst gating are, however, in some decoders derived from the line sync pulses.

How is a decoder using R−Y signal alternate line inversion arranged?

Fig. 41 shows in block diagram form a decoder using this technique. Here the phase inverter, controlled as before by a bistable circuit, is included in the R−Y signal path. In all the decoders so far introduced using this technique tuned R−Y and B−Y chroma signal amplifier stages are included between the delay line/signal separation network and the synchronous demodulators.

Some other alternative techniques are used in this decoder.

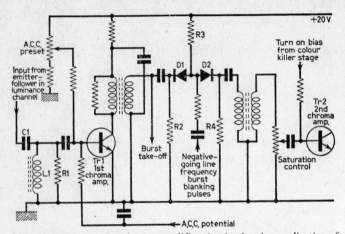

Fig. 42. Representative chroma amplifier circuits showing application of a.c.c., diode gate burst blanking arrangement and colour killer turn on biasing. C1, L1 and R1 form a high-pass filter to remove the luminance signal at the input. The bandwidth of the chroma amplifier stages is much narrower than the i.f. and luminance stages.

First the burst amplifier stage drives two separate detectors, the reference oscillator phase detector and a separate a.c.c. detector, with the colour killer stage operating on the basis of the presence or absence of an a.c.c. potential. And since the a.c.c. potential is derived from the burst amplifier the colour control is here used to vary the gain of this stage. The rest of the decoder operates on the same lines as the previously described arrangements.

How are the chrominance and luminance signals separated?

By including in the input to the first chrominance amplifier a high-pass filter. In the chrominance amplifier circuit shown in Fig. 42 C1, L1 and R1 comprises the high-pass filter.

How is burst blanking carried out?

A diode gate circuit switched off by line frequency pulses derived from the line output stage is commonly used to remove the burst signal from the chroma channel. In Fig. 42 a two-diode (D1, D2) gate circuit is incorporated between the first and second chroma amplifiers. Both diodes are normally biased on by the biasing network R2, R3, R4 so that the signal is passed from the first to the second chroma stage. Negative-going line frequency pulses however are fed to the anodes of the two diodes so that they are biased off during the duration of these pulses, which are timed to coincide with the burst signal, thus removing this from the chrominance channel. The blanking pulses may alternatively be used to switch off one of the chroma stages. In one decoder the pulses are used to introduce heavy damping of the tuned circuit of the second chroma amplifier stage.

In the chroma channel shown in Fig. 42 the burst signal is taken off in the collector circuit of the first chroma amplifier, to the base of which the a.c.c., derived from the rectified and smoothed burst signal, is applied. A preset control in the base circuit establishes the level at which a.c.c. action commences. The saturation or colour control is simply a potentiometer determining the amount of chroma signal tapped off and fed to the second chroma stage. The colour killer potential acts as a turn on bias for the base of the second chroma stage, which is thus cut off in the absence of this bias. The colour killer potential must be applied to a stage following the burst take-off point since it is dependent on the presence of the burst signal. A number of different saturation control arrangements is in use, a common alternative technique being to include the saturation control in the emitter circuit of one of the chroma amplifiers.

How does the delay line circuit operate?

To distinguish the subcarrier-based chroma signals proper from the colour-difference signals, which are at video-frequency, contemporary texts commonly refer to the R−Y chroma signal as V chroma signal, or merely to V signal, and to the B−Y chroma signal as U chroma signal, or merely U signal, so we shall for the purpose of this description at least refer to the V and U notations.

Fig. 43 (right). How the delay line with adder and subtractor networks separates the V and U signals and carries out averaging over successive pairs of lines in the PAL system to overcome spurious phase changes. Separating the V and U signals prior to detection makes the design of the detectors much less critical.

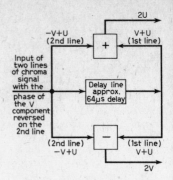

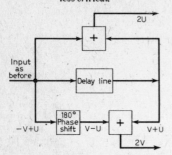

Fig. 44 (left). The subtraction process consists of changing the phase of one set of direct (or delayed) signals by 180° and then adding them to obtain 2V. The V signal is ± 2V on successive lines due to the PAL alternate line V phase inversion.

In Fig. 43 is shown schematically what happens in the delay line circuit with its associated adder and subtractor networks. The inputs fed to the adder and subtractor networks consist of information from two successive lines of picture, with the phase of the V signal reversed on alternate lines. The result of this is that the V components of the signal cancel out in the "adder", whose output consists of the combined U signal from two successive lines, while the U components of the signal cancel out in the "subtractor", whose output consists of the combined V signal from two successive lines, with the phasing of the signals arranged, as we shall see, so that +2V is obtained on one line, while on the next line the output is −2V. Thus the averaging process between lines that is a feature of the PAL system is carried out and the U and V signals separated in one fairly simple operation.

The process is not quite as simple as suggested by Fig. 43, however, because to subtract two signals what needs to happen is a 180° shift in phase of one of them and then addition. This

76

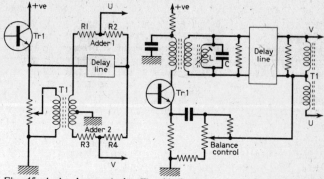

Fig. 45. A simple practical means of carrying out the processes shown in Figs. 43 and 44.

Fig. 46. Commonly used circuit in which to equalise the gain of the direct and delayed signals the delay line is driven from the collector of the driver stage while the direct signal is tapped off at its emitter. Both Tr1 and T1 provide 180° phase shifts. The tuned circuit L, C provides fine adjustment of the delay time.

is shown in Fig. 44. The 180° phase shifting network might be included in the input from the chroma amplifier to the V adder or in the input from the delay line to this network.

A simple method of performing these operations is shown in Fig. 45. The chroma signals are taken from an emitter-follower stage and then fed to the delay line and, via transformer T1, to the two adding networks R1/R2 and R3/R4. The centre-tapped secondary of T1 yields output signals with phase differences of 180° between them, thereby providing the necessary phase shift as shown in Fig. 44, so that from one adder emanates the U signal and from the other the V signal.

In a practical circuit account has to be taken of the 180° phase shift applied to the delayed signal by the delay line (see page 80. Another complication that arises in practice is the attenuation introduced by the delay line, which makes it necessary either to introduce compensating attenuation in the direct signal path or a stage of amplification in the delay line signal path to make good the attenuation.

A commonly adopted circuit is given in Fig. 46, where the gain of the direct and delayed signal paths is equalised by feeding the delay line from the collector of the delay line driver stage Tr1, while the direct signal is obtained from the emitter. Phase shifts of 180° result from Tr1 (base to collector), from the opposite ends of the autotransformer T1 relative to the centre tap and from the delay line itself. Tuned circuit L/C is included to provide fine adjustment of the delay time and hence the signal phasing.

The fact that this circuit is widely used merits a full account of its operation. Let us take the case where the line being received has positive V and U components. This means, then, that U+jV signal is being applied to Tr1 base (the +j notation merely signifies that the vector is at +90° to the X axis while, conversely, −j signifies that it is −90° to the X axis). Since the phase of the signal at the output of an emitter-follower is the same as that of the signal at the input (i.e. no phase change), the signal at Tr1 emitter is thus at that instant also U+jV, which is fed, via the "balance" control preset, to the centre tap of T1, which is an autotransformer.

Now, the delayed signal is that from the preceding line (i.e. U−jV). Since this undergoes two phase reversals, 180° from base to *collector* or Tr1 and 180° in the delay line, it will emerge from the line as U−jV, and from here is connected to T1. Since the opposite ends of T1 are in antiphase relative to the centre tap, the required phasing for addition and subtraction is achieved. To see how this happens let us now consider two successive lines of chroma signal applied to Tr1 base from the chroma bandpass amplifier—a non-inverted line and a following inverted line, corresponding to U+jV and U−jV. These signals thus appear at T1 centre tap. The delay line is also yielding signals of the same phase (owing to the two 180° phase shifts) but one line late, so the delay line delivers U−jV and U+jV respectively. At the top of T1, therefore, we get net addition such that (U+jV)+(U−jV)=2U on one line and (U−jV)+(U+jV)=2U on the next line, while at the bottom, owing to the phase reversal resulting from the autotransformer action, we get addition again but this time at the opposite phase such that $[-(U+jV)+(U-jV)]=-2jV$ on one line and $[-(U-jV)+(U+jV)]=+2jV$ on the next line. In other words, the chroma signal fed to the U synchronous detector is devoid of V chroma signal, while that fed to the V synchronous detector is devoid of U chroma signal. It is in this way that the PAL system eliminates "phase sensitivity".

It is worth noting that while the above description refers to two summation networks, the same net effect would be produced by considering the network which yields the V chroma signal as a subtractor (that is, as a subtractor resulting from the integrated functions of phase inversion and addition). The 2U signals are computed as before, but when the conception of subtraction is considered the $+2jV$ and $-2jV$ signals are derived from the expressions $(U+jV)-(U-jV)=+2jV$ and $(U-jV)-(U+jV)=-2jV$, where in both cases one term is subtracted from the other instead of the two terms in both cases being added together as in the former example.

Various circuit artifices have been evolved to provide the PAL action, and the net result of all of them, using a PAL delay line, is the same. It should also be understood, of course, that to cater for the $180°$ phase shift resulting from signal passing through the PAL delay line, as explained in a later answer, either the emerging inverted delayed signal must be phase shifted by $180°$ and then added to the direct signal or the direct signal must be inverted prior to application to the PAL delay line, as in Fig. 46.

What is the effect of phase error on the display from a PAL delay line receiver?

Hue error due to phase error is, in effect, converted to a reduction in colour saturation after demodulation. This results from a reduction in amplitude of V and U signal applied to the demodulators or synchronous detectors, and since, in a correctly adjusted receiver, the amplitude reduction is the same at both detectors for a given phase error the hue remains constant but is reduced in saturation. The saturation reduction is small for the phase error likely to be encountered in practice.

How does the delay line work?

A glass delay line ground to a very precise length and operating on ultrasonic principles is generally used to provide the required line period delay time. Fig. 43 shows the principle of operation. The input signal is applied to a transducer which converts it to an equivalent acoustic wave. This then travels towards the end of the delay line, where it is reflected back to a second transducer which converts the signal back to an electrical one delayed in time by the time taken by the acoustic wave in travelling to the end of the line and back.

Recent lines work with multiple reflections so that the correct path length is achieved within smaller overall dimensions.

Why is the delay time 63·943 μs?

Although the line period of the 625-line system is 64μs, the PAL delay line introduces a delay of 63·934μs. This is because for the addition and subtraction of direct and delayed signal lines the delay must correspond to an exact number of half cycles of the chroma signal. Since the chroma signal is based on a subcarrier of 4·43361875MHz, approximately 283·75 cycles of subcarrier occur in 64μs, failing to provide the exact half-cycle condition. However, the requirement is met by the 63·934μs delay because this corresponds to the duration of exactly 283·5 subcarrier cycles. It will be understood that the odd quarter cycle resulting from a 64μs delay would interfere with the correct phase relationship of the signals and thus detract from the correct PAL action.

An interesting by-product of the 63·934 μs delay, resulting from the odd half-cycle multiple, is that the line yields a delayed signal which is 180° out of phase with the direct signal which, of course, needs to be taken into account in the PAL decoder and when studying the PAL decoder action in detail. We have already seen in a previous answer that the decoder design shifts by 180° either the inverted and delayed signal emanating from the PAL delay line before it is added to the direct signal or the direct signal before it is applied to the delay line as, for example, in Fig. 46.

Since the delay is not exactly equal to the time taken by the video signal of one line, does this not affect the display?

There is a very small departure from absolute colour/luminance registration in the combination of two successive lines, but since this is outside the colour resolution of the human eye at normal viewing distance it is of no consequence.

What is meant by "synchronous demodulator"?

A synchronous demodulator provides a demodulated output only when two signals, a reference or control signal and the signal to be demodulated, are applied to it.

How do synchronous demodulators work?

In PAL-D receivers each chroma demodulator receives only one of the two modulation products since, as already explained, the PAL delay line and associated phasing and summation networks resolve the multiplexed chroma information into V and U signals. Thus the V demodulator receives V signal and yields after "de-weighting" the original R−Y signal, while the U demodulator receives U signal and yields again after "de-weighting" the original B−Y signal. The demodulators require the addition of reference signal of correct phase and frequency to make good the real subcarrier which was suppressed at transmission.

It is best to regard the demodulators as "sampling" devices which are synchronised so that each samples the amplitude of the appropriate colour-difference information contained in the V and U signals. The idea is shown in Fig. 48, where at (a) is shown the U signal applied to the U or B−Y demodulator, at (b) the V signal applied to the V or R−Y demodulator and (c) the reference signal. The sampling is controlled by the reference signal, such that the positive peaks of the signal switch on the appropriate demodulator for a small period of time.

Each demodulator has its own reference signal input and to satisfy the quadrature requirement the signals are 90° apart. The reference signal for the V demodulator is shown in dotted

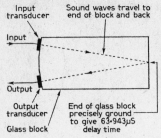

Input transducer

Sound waves travel to end of block and back

Input

Output

Output transducer

Glass block

End of glass block precisely ground to give 63·943µS delay time

Fig. 47 (left). The principle of the ultrasonic glass delay line. The length of the glass block is precisely ground but the sides actually have an irregular surface to prevent interference reflections within the block. More recent lines are designed for multiple reflections so that the correct path length is obtained within smaller overall dimensions.

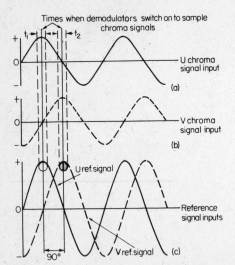

Times when demodulators switch on to sample chroma signals

t_1 t_2

U chroma signal input

(a)

V chroma signal input

(b)

Uref.signal

Reference signal inputs

90°

Vref.signal   (c)

Fig. 48. Principle of synchronous demodulation, showing how the reference signal with 90deg. phase displacement between the two demodulators samples the chroma signals by causing the detectors to switch on for a short time around the peak of each chroma signal cycle. Incorrect phasing will impair the colour operation (see text).

line. Clearly, from this illustration, then, we can see that when the U demodulator is switched on (at t_1) the U chroma signal is at peak amplitude while the V chroma signal is passing through zero, and vice versa. The switching period, t_1 for the U signal and t_2 for the V signal, is sometimes called the "angle of flow" of the demodulator, and can differ widely between circuits, depending on design.

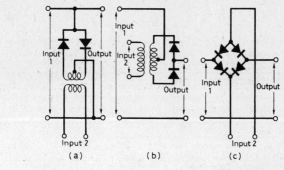

(a)   (b)   (c)

Fig. 49. Three synchronous demodulator circuits. (a) shunt diode pair, (b) series diode pair and (c) series diode bridge.

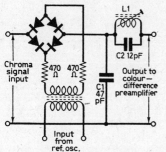

Fig. 50 (left). The widely used series diode bridge synchronous demodulator with low-pass filter in the output.

This illustration reveals the importance of correct phasing of the reference signal and of the correct $90°$ phase difference between the signal at the two demodulators. In a PAL-D receiver, incorrect phasing of switching at both demodulators will result in reduction in saturation, which is the cost of the PAL system correcting the hue in spite of the misphasing, while incorrect phasing of switching at one demodulator only will affect the hue since then the ratio of the demodulated colour-difference signals (R−Y and B−Y) will also be in error. Incorrect phasing of the switching will also introduce crosstalk between the R−Y and B−Y outputs.

There are several ways by which synchronous demodulation can be achieved, and three basic circuits are given in Fig. 49.

The principle in each case is that the demodulator is switched on by the reference signal *once only each reference signal cycle*, the reference signal generally being applied to input 2 and the appropriate chroma signal to input 1. Since the sampling frequency is much faster than the frequency at which the colour changes, the output faithfully follows the colour variations in the original scene. A $90°$ phase-shift network is included in the feed from the reference generator to one demodulator.

The circuits at (a) and (b) are shunt and series diode-pair demodulators respectively, with one input applied via a transformer with a centre-tapped secondary. The circuit at (c) is the widely used series bridge demodulator, where, in Fig. 50 in practical form, L1/C1 form a low-pass filter, tuned by C2, to suppress the reference signal, so that only the colour-difference signals are passed to the following stages. All chroma demodulators incorporate this sort of filtering.

Quite a few modern receivers are now employing a chroma demodulator i.c. which has sections for V detector switching and synchronisation as well as for colour-difference pre-amplification.

How is the G−Y signal recreated?

The G−Y colour-difference signal is recreated using a simple resistor matrix network of the type already referred to. This matrixing may be carried out in the colour-difference preamplifier stages, in the colour-difference output stages or in an integrated circuit.

In Fig. 51 the R−Y and B−Y outputs from the synchronous demodulators are fed to the bases of the R−Y and B−Y preamplifier stages Tr1 and Tr3. From the collectors of these two common-emitter stages −(R−Y) and −(B−Y) drive signals are obtained, since there is phase reversal in a common-emitter stage, to drive the R−Y and B−Y colour-difference output stages. R−Y and B−Y signals appear at the emitters of Tr1 and Tr3, and the result of matrixing these signals in the correct proportions in the matrix network R1, R2 (with R3 providing

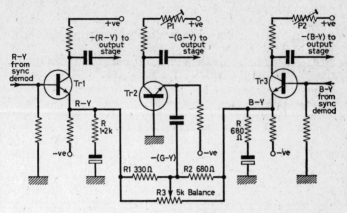

Fig. 51. G—Y matrixing in the emitter circuits of the R—Y and B—Y colour-difference preamplifier stages.

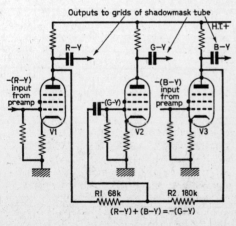

Fig. 52. An alternative G—Y matrixing arrangement—carried out in the anode circuits of the R—Y and B—Y output stages.

85

balance adjustment) is the derivation of a $-(G-Y)$ colour-difference signal. This must be amplified, but without phase inversion since this is the signal required to drive the $G-Y$ colour-difference output stage (to keep the three colour-difference signals in phase with each other). For this reason the $G-Y$ colour-difference preamplifier stage Tr2 is operated in the common-base mode, providing the required $-(G-Y)$ output. The values of the resistors marked R are selected to preset the gain of the $R-Y$ and $B-Y$ stages, the gain required from each channel being different because of the different weighting given, as explained in Section 1, to the $B-Y$ and $R-Y$ signals before modulation at the transmitter. P1 and P2 presets allow for balancing of the gains of Tr2 and Tr3, relative to the gain of Tr1, for optimum colour balance. They are called colour-difference drive controls and are adjusted, in conjunction with the $G-Y$ matrix control R3 if fitted, using a colour-bar test pattern. Adjustments must be made in darkness with correct highlight settings, i.e. first check that the grey-scale tracking (see Section 2) is correct.

The other widely used approach is to carry out the matrixing in the anode circuits of the $R-Y$ and $B-Y$ colour-difference output stages as shown in Fig. 52. Here we have $R-Y$ and $B-Y$ outputs from the $R-Y$ and $B-Y$ colour-difference output stages, and the result of matrixing is to derive a $-(G-Y)$ signal. This is fed to the grid of the $G-Y$ colour-difference output stage to obtain the required $G-Y$ output.

Colour-difference output stages have already been featured in Section 2 (see Fig. 25, for example).

How do the burst stages operate?

The burst take-off point may be at the input to the decoder or following one (as in Fig. 42) or two stages of chroma amplification. Two burst amplifier stages are generally used, and one of these must be gated, i.e. switched on, for only a brief period each line by a pulse timed to coincide with the burst signal (which occurs during the back porch period following the line sync pulse). In this way the chroma information is removed and only the burst signal passed on to the phase detector circuit which is driven by the final burst amplifier stage.

To gate the gated amplifier stage a pulse derived from the line output stage or one based on the sync pulse may be used. In the latter case, and sometimes in the former, the pulse is used to excite a ringing circuit. This technique is used in the burst channel shown in Fig. 53, where Tr1 is the first burst amplifier and Tr2 the gated burst amplifier. The negative-going sync pulse is applied to the transformer T1, which then rings: the first ring, i.e. positive-going overswing, occurs during the back porch period when the burst signal is present, and this positive overswing is fed via R1 and C1 to Tr2 base so that it is switched on and passes the burst signal to the phase detector transformer T2.

When, instead, positive-going line flyback pulses are used for gating, they are shaped and applied via a delay network so that it is the positive tip of the pulse that provides the gating action (assuming an npn-gated stage).

How does the phase detector circuit operate?

The phase detector input transformer has a centre-tapped secondary the ends of which feed a phase-sensitive rectifier

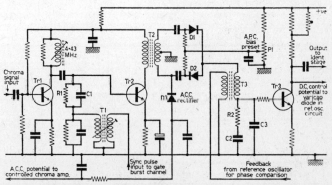

Fig. 53. Typical burst amplifier stages and phase detector circuit. This part of the circuit also generally incorporates the a.c.c. rectifier as shown.

87

consisting of two diodes D1 and D2 (Fig. 53). The operation of the circuit is similar to that of flywheel line sync circuits in monochrome receivers. The control potential is established at the junction of the two diodes, and fed via the d.c. amplifier stage Tr3 to the oscillator to lock it in phase to the burst signal. Part of the oscillator output is fed back to transformer T3. If it is out-of-phase with the burst signal the detector circuit is unbalanced and the control potential alters so as to bring the oscillator back into phase with the burst signal. When the phase of the oscillator is in advance of the burst signal the d.c. control potential increases positively; when it lags the burst signal the control potential decreases. The control potential is used to vary the capacitance of a reverse-biased variable-capacitance diode in the oscillator tuned circuit. The usual flywheel network (R2, C2, C3) is included in the base circuit of the d.c. amplifier to give a long time constant to the control loop.

The type of phase-sensitive detector shown in Fig. 53 is very widely used. An alternative used in one decoder is a bridge diode detector circuit.

How does the ident signal arise?

The burst signal swings $\pm 45°$ about the $-(B-Y)$ axis because of the inversion of the $R-Y$ component of the composite chroma signal on alternate lines. This gives rise to a $7 \cdot 8$ kHz ripple signal—the ident signal—in the phase-detector circuit. This may be extracted at the output of the phase detector or from the d.c. amplifier as in Fig. 53. As the time constant of the oscillator control circuit is long compared with the ident signal, the ident signal does not interfere with the d.c. conditions of the control circuit.

What a.c.c. techniques are in use?

The general arrangement is that the burst signal is rectified and then filtered, the resulting smoothed d.c. voltage then constituting the control potential, as shown in Fig. 53. This potential is used as bias for the controlled chroma stage, as in Fig. 42, such that as the potential falls so the gain of the stage increases, and vice versa. Since the amplitude of the chroma signal is reflected in terms of burst signal amplitude, this technique ensures that the chroma signal fed to the PAL delay line and summation networks and thence to the synchronous detectors holds at a constant amplitude irrespective of changes in signal conditions outside or inside the receiver.

Various schemes for manual gain control of the chroma channel are in use, some by straight bias/gain control and others by gain control due to impedance change or even by change in signal conductivity of a diode circuit.

What form does the ident stage take?

The ident stage consists of a tuned amplifier having a high-Q circuit resonant at the ident signal frequency of 7·8 kHz. This stage is fed with the 7·8 kHz ripple signal, taken either direct from the output or the phase detector circuit or, as shown in Fig. 53, from a following d.c. amplifier stage. The output of the ident stage is used to control the switching of the R−Y alternate line phase inverter to keep it in synchronism with the switching at the transmitter and, in most decoders, as the basis of the colour killer system. In addition in one decoder the ident signal is rectified to provide the a.c.c. potential, providing a noise-free source of a.c.c.

A representative circuit is shown in Fig. 54. The 7·8 kHz ripple is applied to the base of the ident amplifier Tr1 in the collector of which the 7·8 kHz tuned circuit is connected. This particular circuit operates in conjunction with an emitter-follower stage Tr2. In the absence of the ident signal this stage is cut off since no fixed base bias is applied to it. When the ident signal appears at Tr1 collector, however, it is rectified by D1 to provide a base bias for Tr2. As a result the 7·8 kHz signal appears at Tr2 emitter and is fed back via R1 to the 7·8 kHz tuned circuit, providing in this way positive feedback to increase the amplitude of the ident signal. The 7·8 kHz output at Tr2 emitter is fed via C2, R2 and D2 to the R−Y alternate line inverter stage to control its operation and it is also rectified by the colour killer rectifier D3, smoothed and used as a turn on bias for one of the chroma amplifier stages (see, for example, Fig. 42).

Are there alternative methods of colour killing?

The technique so far described and illustrated in Figs. 42 and 54 is in very wide use. However there are other possibilities. For example, the presence of the a.c.c. potential may be used as the basis of colour-killing operation. Another approach is illustrated in Fig. 55. Here again the ident signal is the basis of the action. When it is present it is rectified by D1 and drives Tr1 into conduction. As a result the collector voltage of Tr1, a pnp transistor, falls from a high negative value to almost chassis potential and the chroma stage Tr2, an npn transistor to the base of which Tr1 is linked by R1, is switched on. In the absence of the ident signal Tr1 is non-conducting and its collector voltage rises to a high negative value, switching Tr2 off.

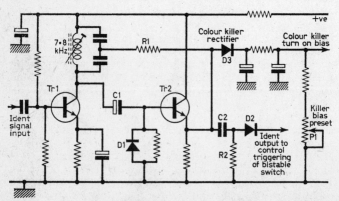

Fig. 54. Typical ident amplifier stage with emitter-follower (Tr2). Two outputs are provided, one via D2 to control the triggering of the bistable switch and the other a colour killer turn on bias from D3 via a filter network.

*Fig. 55. An alternative method of pro-
viding a colour killer bias.*

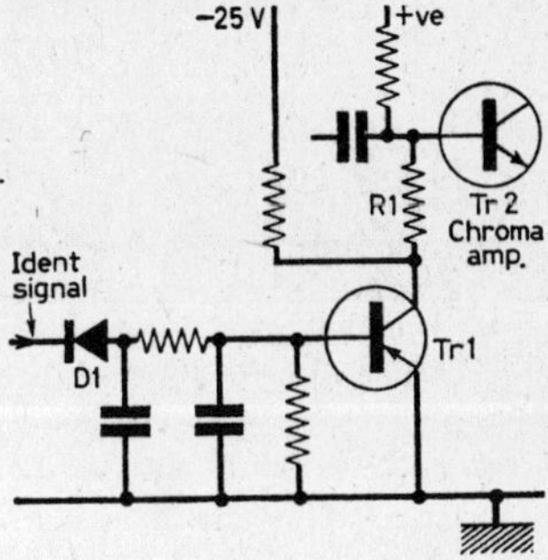

What type of reference oscillator is used?

The reference oscillator output is used to control the
synchronous demodulators, and to provide accurate detection
of the quadrature modulated signal must be very stable in
operation. For this reason a crystal oscillator circuit is always
used, and is controlled, as we have seen, by a feedback phase
control system. It is also the general practice to incorporate
a buffer stage between the oscillator and its load. A
Colpitts-type oscillator with feedback from base to emitter
is favoured, a typical example being shown in Fig. 56. The
control potential from the phase detector or d.c. amplifier is
applied, as shown, as a reverse bias to the cathode of the
varicap diode D1, varying the capacitance of the diode to lock
the phase of the oscillator to that of the transmitted burst
signal. The emitter-follower Tr2 acts as the buffer stage, its
output being fed back to the phase detector circuit and also to
the synchronous demodulators, via a 90° phase shift network
in the case of the B−Y demodulator and, in decoders where
the phase of the oscillator signal is reversed on alternate lines,
to the R−Y alternate line phase inverter in the case of the
R−Y demodulator.

It will be appreciated that not only must the reference
generator produce a signal of exactly the correct frequency
but the signal must also be as close as possible to the phase
reference of the original (suppressed) subcarrier. If there is a

91

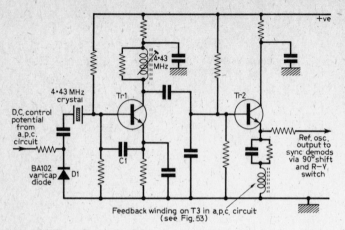

Fig. 56. Typical crystal reference oscillator with emitter-follower buffer stage (Tr2).

slight phase difference between the two signals, then this must be very small and of a constant value to allow the synchronous chroma demodulators to switch accurately once every signal cycle. Minor phase error, provided it is of a constant value, merely has the effect of reducing the saturation in PAL-D receivers, as we have seen, and this can be easily corrected by slightly advancing the colour or saturation control on the receiver.

What does the term passive subcarrier generator mean?

This refers to a technique which derives the reference signal or subcarrier direct from the bursts instead of from an oscillator (i.e. active circuit) controlled by an a.p.c. circuit. The scheme incorporates an electronic switch in the burst circuit to eliminate the PAL line-by-line phase changes so as to make the bursts of constant phase. The constant phase signal is then itself shifted in phase to the angle required to line up with the V and U axes of the synchronous demodulators. The electronic switch may be a simple diode circuit operated by the bistable

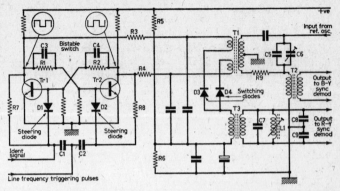

Fig. 57. Bistable circuit (Tr1, Tr2) controlling the R−Y alternate line phase inversion system (T1, D3, D4).

multivibrator which is already in use for the V detector phase reversing on alternate lines (also see later).

The resulting signal is at the subcarrier frequency (4·43361875 MHz) but 100 per cent modulated by a square wave of line repetition frequency, the mark/space ratio being about 28:1. The sidebands resulting from the modulation are removed by passing the signal through a very high Q filter consisting of a crystal (such as that used in active reference generator circuits), the signal emerging being of continuous sinewave form at the subcarrier frequency. Before application to the V and U synchronous detectors, the signal is amplified and limited so that it has a constant amplitude between bursts and is of suitable level to drive the synchronous detectors.

The technique may also be exploited for colour killing, since the subcarrier occurs only when bursts are present thereby allowing a bias to be derived from the subcarrier to operate the colour killer circuit, and for synchronisation of the V detector phase switching (ident, see later) owing to the fact that the electronic switch which removes the PAL swings from the bursts also operates the V detector phase switching, and when the switch is operating on the incorrect count the subcarrier signal falls to zero. Integrated circuits providing all these functions are used in the new generation of colour receivers.

What form does the 90° phase shift network take?

There are several ways in which the necessary 90° phase shift between the feeds to the two synchronous demodulators may be obtained. Use may be made of the fact that between two loosely coupled resonant circuits the voltage in the secondary will lag by 90° the voltage in the driven primary; or of the fact that when an r.f. signal is applied to a resistor and reactance connected in series the voltage at the junction will lead or lag in phase the applied voltage depending on whether the reactance is inductive or capacitive; or a small delay line may be used to provide the 90° phase shift. All these techniques are in use. In the circuit shown in Fig. 57 the reference oscillator signal is fed via T1 to transformer T2 from which the feed to the B−Y synchronous demodulator is taken. The *RC* network R9, C5, C6 provides the 90° phase shift, C6 enabling fine adjustment to be made.

How is the R-Y signal alternate line phase reversal achieved?

In the example shown in Fig. 57 the technique of reversing the phase of the reference oscillator signal to the R−Y synchronous demodulator on alternate lines is used. T3 feeds the R−Y synchronous demodulator, its input being obtained from T1 via either diode D3 or D4 depending on which diode is conducting. The two secondary windings on T1, the primary of which is driven by the signal from the reference oscillator, are wound so as to provide anti-phase outputs. Thus with diodes D3 and D4 fed from these two windings and switched on and off alternately at line frequency, the phase of the reference oscillator signal to the R−Y synchronous demodulator is reversed on alternate lines.

How are the diodes switched on and off?

This is where the bistable circuit Tr1, Tr2 comes in. This is a member of the multivibrator family of circuits and has two stable states, either Tr1 fully conducting and Tr2 cut off or

Tr2 fully conducting and Tr1 cut off. Since the collector voltage of each transistor will be very low when it is bottomed, i.e. fully conducting, and will rise to almost the supply line potential when it is cut off, and since the reversal of the two states, from Tr1 cut off and Tr2 fully conducting to Tr1 fully conducting and Tr2 cut off, occurs very rapidly, the two transistors provide anti-phase square wave outputs, as shown, which are used to bias D3 and D4 on and off alternately at line frequency. Tr1 and Tr2 themselves are switched on and off by means of line frequency pulses fed to their bases via capacitors C1 and C2 and pulse steering diodes D1 and D2.

How does the bistable circuit function?

Let us suppose that Tr1 is fully conducting. Its collector voltage will fall to almost chassis potential, and this voltage will appear at Tr2 base because of the coupling resistor R1, cutting Tr2 off since it then has almost no base bias. In this condition Tr2 collector voltage will rise almost to the positive supply potential, and because of the resistive coupling to Tr1 base via R2 this positive potential will appear at Tr1 base maintaining Tr1 in the fully conducting state. This situation is stable and continues until some external circumstance occurs to alter it. The external circumstance in this case is the arrival of line frequency pulses at the junction of C1 and C2. These could be negative- or positive-going and could be used to switch on the non-conducting transistor or to switch off the conducting transistor. The usual technique is to use negative-going pulses to switch off the conducting transistor. Suppose that Tr1 is the transistor that is conducting. The negative pulse applied to its base will cut it off, its collector voltage will rise, and the increased voltage appearing at Tr2 base will make it conduct. Tr2 collector voltage will then fall, holding Tr1 in the cut off condition. This situation continues until the arrival of the next trigger pulse, which is fed to Tr2 base to cut if off.

C3 and C4 are sometimes fitted, sometimes not. Their function is to speed up the switching action between the two stable states of the bistable circuit. They are of small value—about 470pF—and have a differentiating effect on the waveforms fed via the collector-base cross-coupling resistors R1 and R2. They are called speed-up capacitors for this reason.

Resistors R3 and R4 limit the current used to switch on the switching diodes D3 and D4. A small standing bias is applied to these two diodes from the potential divider R5, R6.

A filter circuit, C7, C8, C9, L1 in the example shown, is sometimes included in the feed to the R−Y synchronous demodulator in this type of circuit to remove switching transients.

What is the function of the steering diodes?

The steering diodes are included to speed up the switching action, their function being to steer the input pulse to the transistor next to be switched. Both diodes are reverse biased, but the bias applied to each depends on the state of the associated transistor (the anode of each diode is connected to the base of the associated transistor and its cathode returned via R7 and R8 to the collectors) so that one diode will have a substantial reverse bias while the other has only a slight reverse bias. Suppose, for example, that Tr1 is conducting. As its collector voltage is low only a small positive potential is applied to D1 cathode. On the other hand Tr2 is cut off and a substantial positive reverse bias is applied to D2 cathode. Thus the negative switching pulse sees on arrival a low-resistance path via D1 and a high-resistance path via D2, and is thus steered via D1 to Tr1 base to cut it off.

How is the ident signal used to synchronise the R−Y switching?

The ident signal is, as shown in Fig. 57, applied to one side of the pulse input circuit of the bistable switch. As we have seen the ident signal swings positively and negatively at half-line frequency. If the trigger pulse applied to Tr1 base is in synchronism with the R−Y alternate line switching at the transmitter, the ident signal has no effect. If, however, the trigger pulse is out-of-phase with the transmitter R Y switching the ident signal blocks the trigger pulse (the positive excursion of the ident signal being sufficient to suppress the

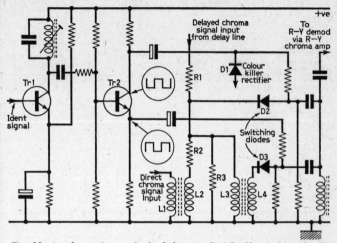

Fig. 58. An alternative method of alternate line R−Y switching. In this instance the switching is done in the chroma signal path. R1, R2, R3 comprise the R−Y adder matrix.

trigger pulse) and the bistable circuit remains in its stable state until the next trigger pulse arrives. This then passes through to trigger the bistable circuit. In practice the negative-going excursion of the ident signal is generally removed (by diode D2 in the circuit shown in Fig. 54). The ident signal has no further action since the switching remains in synchronism until the set is switched off.

Are there alternative methods of carrying out R−Y switching?

While the scheme outlined above is very widely used, sometimes however in the chroma signal input to the R−Y synchronous demodulator instead of in the reference oscillator signal path, some decoders make use of the ident signal to carry out the alternate line R−Y phase inversion, again in conjunction with a pair of switching diodes. An example with the switch in the chroma signal path is shown in Fig. 58. The second indent stage Tr1 feeds a squarer phase inverter stage Tr2 which provides, as shown, at its collector and emitter anti-phase square wave outputs, these being squared versions of the sinewave ident signal applied to its base. These square wave outputs are used to switch on alternately line by line the switching diodes D2 and D3 so that the signal passed to the R−Y synchronous demodulator via the R−Y chroma amplifier is from D2 on one line and from D3 on the next line.

As the switching takes place in the R−Y chroma signal path instead of in the reference oscillator signal path Fig. 58 also shows the R−Y matrixing circuits used in this decoder. The direct signal from the chroma amplifier stage is fed via coils L1 and L2 which provide the $180°$ phase reversal required to give signal subtraction to one end of the matrix network R1, R2 and R3. The delayed signal from the 64 μs delay line is applied to the other end of the matrix. Thus at the junction of R1, R2 we have across R3 $2(R−Y)$ and $−2(R−Y)$ on alternate lines. When D2 is switched on it passes the $2(R−Y)$ signal to the following stage. When D3 is switched on it receives its input from the matrix via the $180°$ phase shift network L3, L4 so that it also passes a $2(R−Y)$ signal to the following stage.

Since the alternate line switching is controlled directly by the ident signal it is automatically synchronised to the alternate line R−Y switching at the transmitter.

In subsequent production this circuit was modified, using a single-ended arrangement, but still operating on the same principle with the diodes D2 and D3 alternately switched on so that L3−L4 provide the required R−Y phase inversion on alternate lines.

A further technique found in some early decoders is the use of a ring modulator circuit to control the line by line phase inversion instead of a pair of switching diodes as used in the examples illustrated in this section. The ring modulator in one decoder is directly controlled by the ident stage whilst in another it is switched by a bistable circuit.

To what extent are integrated circuits now used in colour receivers?

Some of the very recent receivers (at the time of writing)

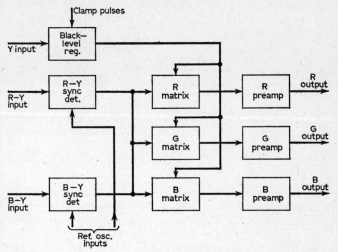

Fig. 59. The main operations performed by the integrated circuit, which provides R, G and B outputs, featured in some Rank-Bush-Murphy colour models.

employ integrated circuits (i.c.s) in the video processing and line oscillator stages as well as in the intercarrier sound stages and in the decoder sections. Indeed, as many as six i.c.s may be used for such functions as i.f. gain, video detection, intercarrier sound, line oscillator/sync separation, PAL decoding and chroma demodulation/matrixing for RGB drive. The use of i.c.s significantly reduces the number of separate components and transistors, thereby greatly simplifying the design, layout and manufacture of the receivers, Fig. 59 is one example.

Could you give an example of an i.c. decoder arrangement?

A block diagram of a decoder using the Mullard TBA series of i.c.s is given in Fig. 60. Here the luminance or Y signal is processed by the TBA500Q, the chroma signal by the TBA510Q, which receives an input from the luminance i.c., and the chroma detection by the TBA520Q or TBA990Q,

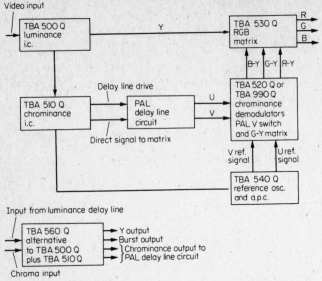

Fig. 60. Example of decoder based on the Mullard TBA series of i.c.s.

which receives V and U chroma signals (i.e. weighted versions of the R−Y and B−Y subchannel information based on the subcarrier frequency) from the PAL delay line circuitry.

The latter i.c. also incorporates sections for PAL V detector switching and G−Y matrixing. For synchronous detection, of course, accurately phased reference signals of subcarrier frequency are required, and these are provided by the TBA540Q reference oscillator and automatic phase control (a.p.c.) i.c., which picks up bursts from the chroma i.c. Thus, from the chroma detectors are yielded the three colour-difference signals (R−Y, G−Y and B−Y). These are matrixed separately with the Y signal at the TBA530Q matrix i.c., which then delivers the three primary colour signals, red, green and blue, which are conveyed to the primary colour output stages and thence to the cathodes of the picture tube.

The functions of the TBA500Q and the TBA510Q are generally combined into a single i.c., such as TBA560Q shown separately at the bottom of Fig. 60.

In general, the internal circuits of the i.c.s follow those based on discrete components, though in many cases there are changes in detail so that the full advantage can be taken of the monolithic i.c. construction in terms of circuit sections. The i.c.s. may contain 70 or so transistors.

How do decoder faults affect the picture?

If the decoder goes completely inactive then, assuming the rest of the receiver is operating correctly, the picture will appear in monochrome.

Loss of colour or drifting bands of colour not properly synchronised to the luminance components of the display can be caused by reference oscillator failure or incorrect adjustment of the oscillator (i.e. P1 in Fig. 53). A fault in the colour killer circuitry or in the burst or chroma stages can also delete the colour, while incorrect (complementary) colours may be displayed if the V detector is switching on the incorrect V signal phase. This could indicate trouble in the bistable or "ident" circuitry.

Some receivers are equipped with a colour killer threshold preset (i.e. P1 in Fig. 54) which, if incorrectly adjusted, may retain colour killer action even when bursts are present. The first step is to disable the colour killer to prove whether the loss of colour is caused by an actual fault (the resulting display then giving some clue as to the whereabouts of the fault) or lack of colour killer switching.

Another decoder fault manifests as Hanover blinds, so called since they were first investigated in Hanover where the PAL system was developed. The symptom arises when the delayed and direct signals in the PAL delay line matrix circuit vary in phase or amplitude with respect to each other. The picture is then affected by horizontal colour striations owing to the eye failing to be completely deceived when the picture elements on successive lines of a field are substantially removed from the true colours as scanned by the camera; the eye then becoming aware of the colour differences along successive lines. The effect is aggravated because a complete frame is composed of two interlaced fields and because the PAL phase alternations are such that the lines containing the errors of one field occur adjacent to the lines containing the same error of the interlaced field.

A receiver suffering from Hanover blinds would either have a definite fault condition in the decoder or be out of adjustment, particularly in the area of the PAL delay line matrixing and direct/delay signal balancing. Preset controls are available to restore the correct balance, but to optimise the adjustments reference should be made to the manufacturer's service manual or similar data (i.e. *Newnes Colour TV Servicing Manual* by Gordon J. King and published by Newnes-Butterworths). Severe blinds occur when the PAL switch fails to operate.

Phasing maladjustment relative to the synchronous demodulators can put the hues in error owing to imbalance of the ratio of the R−Y and B−Y signals delivered by the demodulators when the phase displacement between the reference signal applied to them is not exactly $90°$. On the other hand, when the phases of the reference signal to *both* demodulators are equally in error the ratio of the colour-difference signals is not put in error, which means that the hue will be correct, the effect then being saturation loss due to the phase compensating effect of the PAL system.

Another fault condition shows as pattern interference, called sound/chroma beats, and results from interaction between the sound and chroma signals at the detector. Filtering at the sould i.f. frequency is sometimes adopted to reduce the effect, so if this filter is out of adjustment the interference could be troublesome, as also when the fine tuning of the required channel is not properly optimised. This adjustment should be made for maximum picture definition consistent with minimum intercarrier buzz (from the loudspeaker) and minimum patterning. The beat occurs at 1·57MHz, the difference between the two signals concerned. Misalignment of the i.f. channel is another cause, particularly when the level of the sound carrier at the vision detector is too high (i.e. within 30dB of the vision carrier level). Modern receivers are further minimising the trouble—which is not at all serious, anyway—by the use of a synchronous vision detector (part of an i.c.) in place of the more conventional envelope detector.

The colour performance of a receiver can be appraised by studying the displayed colour bars, which are transmitted from time to time by the broadcasting authorities (between test transmissions and other information) and which are present at the top of the test card itself. Colour faults, of course, can arise from shortcomings in areas other than the decoder. For example, if only the green beam of the tube is operating the display will take on a predominantly green hue, and only the colour bars containing a green component will be displayed. Similarly with either of the other two colours.

It is obviously impossible in the scope of this chapter or, indeed, within the relatively small compass of the whole of this book to investigate all likely faults. A book concentrating more on the subject matter of "servicing" would be necessary, and readers interested in this aspect of colour television may

Colour bar hue	White	Yellow	Cyan	Green	Magenta	Red	Blue	Black
Tube guns on for correct hues	R,G,B	R,G	B,G	G	R,B	R	B	None
Main colour-difference signal(s) present	None	-(B-Y)	-(R-Y)	-(B-Y) -(R-Y)	B-Y R-Y	R-Y	B-Y	None
Blue gun only on	Blue	Black	Blue	Black	Blue	Black	Blue	Black
Red gun only on	Red	Red	Black	Black	Red	Red	Black	Black
Green gun only on	Green	Green	Green	Green	Black	Black	Black	Black
Blue gun off	Yellow	Yellow	Green	Green	Red	Red	Black	Black
Red gun off	Cyan	Green	Cyan	Green	Blue	Black	Blue	Black
Green gun off	Magenta	Red	Blue	Black	Magenta	Red	Blue	Black
No B-Y signal	White	Off-white	Green	Green	Red	Red	Black	Black
No R-Y signal	White	Yellow	Pale mauve	Dark green	Blue	Black	Blue	Black
R-Y signal out of step	White	Off-white	Magenta	Dark orange	Blue	Dark green	Purple	Black

Fig. 61. Showing how the standard colour bars tend to alter in hue under various fault conditions.

find the second edition—containing a colour fault procedure chart—of *Colour Television Servicing*, by Gordon J. King, published by Newnes-Butterworths, useful.

To summarise this short section on colour faults, however, Fig. 61 gives an approximate impression of how the "standard" colour bars tend to change in hue, etc. under various fault conditions.

4

CONVERGENCE

What is the difference between purity and convergence?

The purity adjustment is provided by permanent magnets which affect all three beams in the shadowmask tube, adjusting the angle of approach of the beams to the deflection centre (the centre of the area in which beam deflection takes place) and then to the shadowmask so that at the screen the beams impinge only on the appropriate colour phosphor dots, giving pure colour reproduction without, say, the red beam falling partially on blue dots thus giving incorrect colouring to the picture. The convergence assemblies on the other hand enable the positions of the three beams to be adjusted individually so that the scanning of the three beams can be superimposed to give correct registration of the raster produced by each beam.

What beam movements are required to obtain correct convergence?

Movement of each beam in two directions is required in order to achieve accurate superimposition. Each beam is moved radially with respect to the axis of the tube by the radial convergence assembly, while lateral movement of the beams is provided by the lateral convergence assembly which affects mainly the blue beam and is for this reason usually known as the blue-lateral convergence assembly. The effect of these movements was illustrated in Fig. 34.

How are these movements obtained?

By a combination of permanent and electromagnetic assemblies mounted around the neck of the shadowmask tube. The radial and lateral convergence assemblies have already been illustrated in Figs. 32 and 33 respectively.

What effect do the magnetic fields have on the electron beams?

We must first remember that the electron beams are moving at high velocity from the cathodes towards the screen of the tube under the influence of the high accelerating potentials. The magnetic fields established in the neck of the shadowmask tube by the radial convergence assembly are shown in Fig. 32. Here each magnet assembly, one for each beam, has an associated pair of internal polepieces between which the magnetic fields, which have the direction shown by the broken arrows, are established. Under these conditions the electron beams will be deflected in a direction at right-angles to the direction of the magnetic field by an amount dependent upon the strength of the magnetic field, i.e. radial movements of the beams to an extent dependent on the strengths of the magnetic fields are obtained. Internal screening as shown is included so that the fields do not interact.

The lateral convergence polepieces are all external, the lateral convergence assembly being mounted behind the purity magnets as shown in Fig. 27, and produce fields in the tube as indicated in Fig. 33 by broken lines. The assembly is mounted with the permanent magnet above the blue beam and results in the blue beam being deflected, again in accordance with the strength of the magnetic field, towards the left (looking from the front of the tube) whilst the red and green beams are deflected, to a lesser extent, to the right (though assemblies are available which displace the blue beam only).

What is meant by static convergence?

We have seen that associated with each convergence assembly are permanent magnets which can be adjusted to vary the strength of the magnetic fields. These magnets provide what is

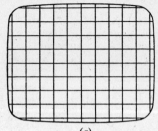

Fig. 62 (a) (Left). Basic cross-hatch pattern, which of course appears on the screen as a white cross-hatch (when the set is correctly adjusted) on a black background. (b) (below). Appearance of the pattern when static convergence (centre of screen) is correct but dynamic convergence is not applied. In practice the red and green horizontal lines may not be accurately converged.

(a)

Yellow line — red and green converged horizontally   Blue line

(b)

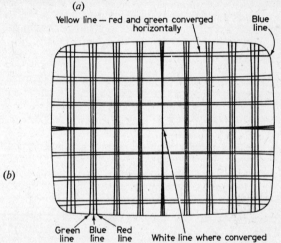

Green line   Blue line   Red line   White line where converged

called static convergence so that the rasters register correctly at the centre of the screen. In some sets this is achieved by feeding a variable d.c. through the radial convergence coils.

What is a cross-hatch pattern?

To enable convergence adjustments to be made a cross-hatch pattern must be available. This type of pattern is shown in Fig. 62(a) and is produced by a cross-hatch pattern generator, an essential piece of equipment in colour receiver ser-

107

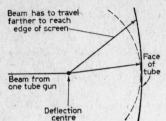

Beam has to travel
farther to reach
edge of screen

Face
of
tube

Beam from
one tube gun

Deflection
centre

Fig. 63. How the distance the beams have to travel from the deflection centre to reach the face of the tube varies over the face of the tube, being greatest at the corners.

vicing. The output from the cross-hatch pattern generator may be at r.f. and injected at the aerial input socket, or at v.f. injected in the v.f. stages, and produces on the screen of the shadowmask tube the type of pattern shown. Fig. 62(b) shows what this pattern looks like when the static convergence is correct but dynamic convergence is not applied. As can be seen the red, green and blue vertical and horizontal lines forming the pattern only coincide at the centre of the screen.

Why does the convergence vary increasingly towards the edges of the picture?

The convergence correction required by the beams changes over the face of the tube because of the tube geometry. This is

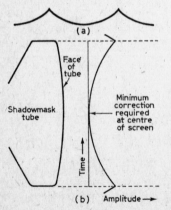

(a)

Face
of
tube

Minimum
correction
required
at centre
of screen

Shadowmask
tube

Time

(b)   Amplitude →

Fig. 64. The basic convergence correction waveform required is a parabola as shown at (a). The reason for this is the geometry of the face of the tube, as shown at (b).

108

illustrated in Fig. 63. If the face of the tube were spherical with respect to the beam deflection centres, the beams would have to travel the same distance from the deflection centres wherever they were deflected to on the tube screen. The static convergence adjustment would then give correct convergence over the entire face of the tube. To give an acceptable viewing picture, however, a much more flat-faced tube is required and this means that, as shown, the beams must travel different distances to reach different points on the face of the tube, the greatest distances being where the beams are deflected to the corners of the picture.

What is meant by dynamic convergence?

Dynamic convergence is provided by the electromagnets on each convergence assembly and is necessary to achieve convergence at the edges of the picture. Clearly the amount of convergence correction required increases with the distance the beams have to travel from their deflection centres to reach the screen so that the required correction varies continuously over the face of the tube. The required correction can only be achieved by means of magnetic fields which are varied at line and field frequency. Hence varying currents at these frequencies are applied to the electromagnets on the convergence assemblies to provide dynamic convergence. The radial convergence assemblies are fed with convergence currents at both line and field frequency: the blue-lateral convergence assembly needs a line frequency convergence current only to give the required dynamic lateral displacement of, mainly, the blue beam.

What shape are these currents?

The basic dynamic convergence current waveform is the parabola as shown in Fig. 64(a). The reason for using this is shown in Fig. 64(b): the waveform is required to provide minimum correction at the centre of the screen where the static convergence magnets establish correct convergence and maximum correction towards the edges of the picture, and this is precisely what a parabolic correction waveform will do.

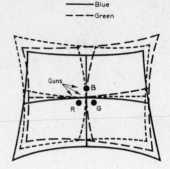

Fig. 65 (right). Without correction the rasters produced by the three guns differ because of the different positions of the guns relative to the axis of the tube. Note the effect on the centre vertical and horizontal lines.

----- Red
——— Blue
— — Green

Guns

B

R   G

Fig. 66 (below). Tilting a parabola. (a) Basic parabolic waveform. (b) Parabola tilted by adding a positive-going sawtooth waveform. (c) Parabola tilted by adding a negative-going sawtooth waveform.

(a)   (b)   (c)

Where are the convergence waveforms obtained?

Since dynamic convergence must take place at the line and field frequencies the correction currents are derived from the timebases.

What is meant by tilting a parabola?

The parabolic convergence waveforms need further shaping before they can be used to give satisfactory dynamic convergence. The need for this is due to the fact that the three guns in the tube are mounted in different positions and as a result trace out slightly different rasters. The effect, in exaggerated form, is illustrated in Fig. 65. To correct this a sawtooth waveform is applied to the parabola and results in the parabola being altered in the manner shown in Fig. 66 depending on whether the sawtooth is (b) negative going or (c) positive going. As can be seen the parabola is tilted, and the process of adding a sawtooth component to a parabolic waveform to adjust the shape of the parabola is called tilting a parabola. For this reason tilt controls are featured on convergence control panels.

Is a parabola always used as the basis of convergence?

Whilst as we have seen the basic convergence correction current required is parabolic, an alternative approach to forming the convergence waveform is possible and is in fact generally used in the case of line convergence. This is to start with a sawtooth waveform and shape this to give the required convergence waveform. The term tilt is in this case also used to describe the shaping adjustment necessary to achieve correct convergence.

Where is the field convergence current obtained?

In sets having a valve-operated field timebase the field convergence current is derived from the cathode of the field output valve. The usual arrangement is shown in Fig. 67. Two coils are wound on each "arm" of the radial convergence assembly, all the coils being connected in parallel.

The waveform available at the cathode of the field output valve is a parabola with a sawtooth component, i.e. it is already a tilted parabola. This is fed via capacitor C1 and the two parabola amplitude controls RV1 and RV2 to the convergence coils. To provide tilt adjustment an opposing sawtooth waveform is obtained from a tertiary winding L3 on the field output transformer and fed to the field convergence coils via the tilt controls RV3 and RV4. This provides variable cancellation of the sawtooth component of the waveform taken from the field output valve cathode. Since the blue gun is the only one central on the screen (see Fig. 65) it requires a slightly different type of convergence waveform and accordingly has its own independent tilt and parabola amplitude controls. To remove the need to readjust the static convergence magnets during or after carrying out dynamic convergence adjustments, d.c. components are added to the field dynamic convergence currents by R2 and R3.

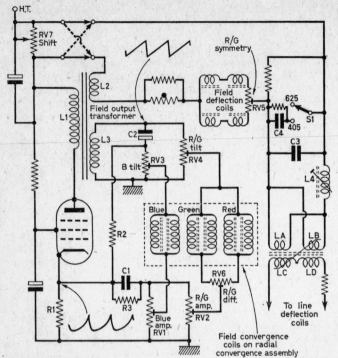

Fig. 67. Field output stage and dynamic convergence circuits; also showing transductor dynamic pin-cushion distortion correction arrangement.

What is meant by matrixed convergence controls?

In theory six controls are required to provide preset dynamic field convergence adjustment: an amplitude and a tilt control for each set of coils, red (R), blue (B) and green (G). In practice it is found that setting up is easier if a different arrangement is used in which the R and G controls are arranged as shown in Fig. 67 where a common R/G tilt control (RV4) and a common R/G amplitude control (RV2) with an

R/G differential control (RV6) are used. These are termed matrixed controls and cause simultaneous horizontal and vertical movements of the R and G beams instead of independent radial movements.

What correction does the R/G field symmetry control provide?

The R/G field symmetry control, RV5 in Fig. 67, enables the field deflection coils to be balanced and is used to equalise the heights of the red and green rasters. It is adjusted in conjunction with the R/G differential control to converge the red and green lines (on a cross-hatch pattern) at the top and bottom of the picture to give yellow lines. Without this symmetry control there is a tendency for the red and green lines at the top and bottom of the picture to be separate though parallel to each other instead of correctly superimposed.

How is field shift carried out?

As we saw in Section 2 shift, or picture centring, cannot be carried out by means of permanent magnets as is the practice in black-and-white only receivers, since this would interfere with the convergence and purity of the picture. Instead a variable d.c. is fed to the field coils. As can be seen in Fig. 67, this is fed to the coils via the d.c. shift control RV7. The associated crossover arrangement determines the direction in which the picture is shifted.

Where are the field convergence currents obtained in transistorised receivers?

In receivers with transistorised field timebases the field scanning current is commonly used as the basis of the field convergence waveform, is tapped off at the vertical R/G symmetry control and is integrated by the field convergence coils.

How is the line radial convergence current obtained?

There are two approaches here: either to integrate the line flyback pulses or to use the line deflection current as the basis of the radial line convergence waveforms. In either case the

113

basic radial line convergence waveform is of primarily sawtooth shape. The latter approach has been adopted as the
general practice in the U.K., the basic radial line convergence
circuit being as shown in outline form in Fig. 68. Here the line
convergence coils are connected in series in the line scan circuit. The amplitude of the waveform in the R and G coils is
preset by RV1 while RV2 presets the amplitude of the B
current waveform. As we saw in the case of field convergence
the shape of the blue radial convergence waveform required
differs from that required for red and green radial convergence, and this means that extra shaping must be provided in
the blue radial convergence circuit to provide a predominantly
parabolic blue convergence current. The components C1, C2
and L1 provide the required shaping, C1 integrating the waveform and C2, L1 providing a second-harmonic (of the line

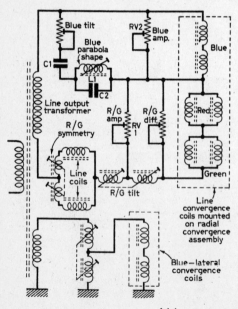

*Fig. 68. Outline cir-
cuit of the line
convergence system
generally used.*

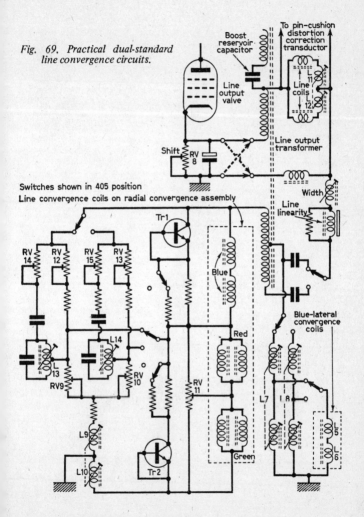

Fig. 69. Practical dual-standard line convergence circuits.

Boost reservoir capacitor

To pin-cushion distortion correction transducer

Line output valve

Line coils
L11
L12

Line output transformer

Shift RV8

Switches shown in 405 position
Line convergence coils on radial convergence assembly

Width

Line linearity

Tr1

RV14 RV12 RV15 RV13

Blue

Red

Blue-lateral convergence coils

L14

L13

RV9 RV10

RV11

L7 L8

L9

L10 Tr2

Green

L5 L6

115

frequency) sinusoidal component. L1 is adjustable and is frequently called the blue parabola control.

How are R/G line tilt and symmetry carried out?

A practical line convergence circuit is shown in Fig. 69, for dual-standard operation. Adding the centre-tapped pair of coils L9 and L10 provides the R/G tilt feature: as the inductance of L9 is increased by adjustment of the core the convergence waveform is given additional differentiation, while an increase in the inductance of L10 provides integration of the waveform.

With convergence applied to a shadowmask tube the horizontal red and green lines tend to cross over as shown in Fig. 70 instead of forming a single converged horizontal yellow line. This is because of magnetic coupling between the field created by the line scan coils and the radial convergence assembly. Correction is provided by means of a line deflection coil balance control—L11, L12 in Fig. 69.

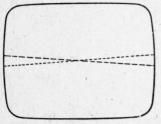

Fig. 70. The red and green horizontal lines tend to cross over as shown here unless a balance control to correct R/G symmetry is included in the line deflection coil circuit.

The line convergence circuits have been complicated by the need for dual-standard operation: as can be seen it has been necessary to duplicate the complete blue convergence waveshaping circuitry and to duplicate some of the other controls. Recent single-standard receivers are free of these extra complications.

It is also necessary to maintain the correct d.c. conditions so that static convergence remains correct on both standards and during and after dynamic convergence adjustment. For this reason clamping is provided by the diode-connected transistors Tr1 and Tr2.

Are the line convergence controls matrixed?

As with the field convergence circuits the red and green controls are matrixed for ease of adjustment. Thus an R/G amplitude control is provided (RV9 and RV10), one for each standard) with an R/G difference control RV11 which acts to increase the current in the R coils whilst decreasing that in the G coils or vice versa. As we have seen a single R/G tilt system is provided.

How is the blue-lateral convergence current obtained?

It has been the general practice in the U.K. to provide dynamic blue-lateral convergence though on many U.S. receivers this is not featured. The lateral convergence current waveform required to provide dynamic blue-lateral convergence is basically a line-frequency sawtooth current of smaller amplitude but similar shape to the line scan current and may be obtained from a winding or other suitable point on the line output transformer. Sometimes a pulse from the line output transformer is used since this, when applied to an inductive circuit, provides a sawtooth waveform. Design details vary considerably from chassis to chassis. In the circuit shown in Fig. 69 the 200 V pulse appearing at the 405 tap on the line output transformer is used, fed to the blue lateral convergence coils L5 and L6. The amplitude of the blue-lateral convergence waveform is preset by means of L7 on 405 lines and L8 on 625 lines.

How is line shift controlled?

Horizontal picture centring is controlled in the same manner as vertical centring, by feeding a variable d.c. to the line deflection coils. In Fig. 69 RV8 controls the amplitude of the d.c. potential to determine the amount of shift applied and a crossover network determines the direction of picture shift.

How is pin-cushion distortion correction applied?

In a black-and-white receiver magnets are used for this purpose. As with picture centring, this cannot be done on colour receivers because the magnets would affect purity and convergence. It is, therefore, general practice to use dynamic raster correction, i.e. overall pin-cushion distortion correction, to overcome raster distortion arising from the fact that the deflection coils are designed for optimum spot quality.

The basic dynamic raster correction technique is to modulate the line scan current with a parabola at field frequency and the field scan current with a parabola at line frequency. A widely used method of doing this is shown in Fig. 67, in which the line and field scanning circuits are linked via the transductor LA—LD. Windings LA and LB on the transductor are connected in series with the field scan coils and control the impedance of coils LC and LD which form a varying load connected across the line scan coils. In this way the line coils are modulated at field frequency since the impedance of the shunt load (LC and LD) across the line coils is varied at field frequency. Conversely line frequency pulses are induced in the control windings LA and LB. These are integrated by C3 and fed to the field scan coils thus modulating the field scanning current at line frequency. L4 presets the amplitude of the line frequency modulation applied to the field coils, with S1 introducing C4 on 405 lines to adjust the time constant of the integrating circuit.

In what order are convergence adjustments carried out?

Before any convergence adjustments are made the set must be degaussed and the normal timebase controls—hold, linearity, height, width, shift, etc.—set correctly. Static convergence adjustments are made first, dynamic adjustments last, and a cross-hatch pattern is required for both. Since the purity and convergence adjustments are interdependent, purity setting is also a part of the complete convergence procedure. The suggested approach is to set the dynamic controls to mid-position (only necessary when complete convergence setting-up is undertaken, for example if a new convergence panel is fitted) then adjust the static convergence magnets, following this with the purity adjustment. The static convergence should then be checked and the magnets readjusted as necessary, and then the dynamic convergence adjustments made. Purity and convergence adjustments must be carried out in darkness and not started till 15 minutes or so after the set has been switched on to allow the circuits to reach normal operating temperature and stabilise.

When is manual degaussing needed?

Automatic degaussing as described in Section 2 affects only the shadowmask, its magnetic shield and protective rim band. Other parts of the chassis and nearby metal objects, e.g. radiators, may however be magnetised and if difficulty is experienced obtaining good purity manual degaussing should be undertaken. This is done with a degaussing coil: suitable for the purpose is a coil of 840 turns of 22 s.w.g. enamelled copper wire wound on a former of about 12 in. diameter. It is convenient to provide the coil with its own on-off switch. The coil is connected to the a.c. mains supply and with the set switched on moved in a circular fashion over the face of the tube and around the front edge of the cabinet, then around the top, bottom and sides (not far back however as the purity magnets and deflection components must be avoided) for about 10 seconds. The corners of the tube are particularly important. Withdraw the coil slowly some eight feet from the set before switching it off. Follow a similar procedure for demagnetising radiators, etc.

How are the static convergence magnets adjusted?

As we have seen four magnets are involved, three radial and one lateral, and all are adjusted by being rotated. Fig. 34 in Section 2 gave an impression of the process of converging the three beams. The red and green beams are first superimposed at the centre of the picture by means of the red and green static radial convergence magnets. The blue radial static convergence magnet is then adjusted so that the blue beam is on the same horizontal axis as the red and green beams, and finally horizontal displacement of the beams is achieved by adjusting the blue lateral convergence magnet to superimpose all three beams. Slight readjustment of the red and green radial convergence magnets may then be necessary. After these adjustments the horizontal and vertical lines at the centre of the picture should be converged as shown in Fig. 62(b). The convergence should be correct on both line standards.

How is the purity adjustment made?

As we saw in Section 2 the purity magnets control colour purity towards the centre of the screen, the deflection coils controlling colour purity towards the edges of the picture. A blank raster is required for purity adjustment, and the blue and green gun first anode drive controls (the picture tube first anode presets, P4, P5 in Fig. 21) should first be set at minimum and the red gun first anode control (P3) at maximum or the blue and green guns switched off where individual switches are provided (S2 and S3, Fig. 21). Then loosen the deflection coils and withdraw them back along the neck of the tube towards the tube base. Colour patches will be seen on the screen, with a large red patch towards the centre of the screen. Adjust the purity magnets by rotation to obtain the largest red area possible central on the screen. Then move the deflection coils forward to obtain an overall red raster. Check the blue and green purity by resetting the first anode preset controls or switches so as to obtain blue and green rasters. Some further slight readjustment of the purity magnets and deflection coils may be necessary. The final check of purity is with all beams on when a pure white raster should be obtained. The picture shift controls may need readjustment after carrying out the purity adjustment, and the static convergence will need further adjustment.

How is dynamic convergence carried out?

Considerable variation exists in the design of dynamic convergence circuits, and it is not possible therefore to give a standard procedure. On the basis of the circuits described in this section, however, the procedure would be along the following lines (component reference numbers refer to Figs. 67 and 69).

First cut off the blue gun and converge the red and green rasters to produce a yellow cross-hatch pattern. The R/G field

120

amplitude (often labelled parabola) control RV2 should be adjusted to remove bowing of vertical lines at the bottom of the picture, the R/G field tilt control RV4 to remove bowing of vertical lines at the top of the picture, the R/G field differential control RV6 to close up the horizontal lines at the top of the picture and the R/G field symmetry control RV5 to close up horizontal lines overall. RV6 and RV5 should be adjusted alternately for optimum results. The R/G line amplitude (or parabola) controls RV9 and RV10 (both systems) are then set to remove left-hand side separation of vertical lines, the R/G line tilt control L9/L10 to remove right-hand side separation of vertical lines, the R/G line differential control RV11 to remove bowing of the horizontal lines at the top and bottom of the picture and R/G line symmetry control L11/L12 to remove crossover of horizontal lines. Adjust the R/G line differential and symmetry controls alternately until best results are obtained.

The blue gun is then turned on and the blue convergence controls adjusted to converge the blue and yellow lines to give a white cross-hatch pattern. The blue field tilt control RV3 removes separation of the blue horizontal lines at the bottom of the picture, the blue field amplitude (parabola) control RV1 removing separation of the blue horizontal lines at the top of the picture. The blue line amplitude controls RV12 and RV13 on 405 and 625 lines respectively are adjusted to remove drooping of blue horizontal lines, the blue line tilt controls RV14 and RV15 being adjusted to remove crossover of the blue horizontal lines. Alternately adjust the blue line amplitude and tilt controls for best results. The blue line parabola controls L13 and L14 are adjusted to remove undulation on the blue horizontal lines and the blue lateral controls L7 and L8 to remove separation of the blue vertical lines at the left- and right-hand sides of the picture.

The dynamic convergence errors corrected in this way are illustrated in Fig. 71. The errors have been exaggerated to emphasise what to look for: the small arrows indicate the correction necessary. After completing the dynamic convergence procedure, check and if necessary readjust the purity;

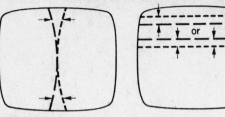

Centre vertical red and green lines bowed at top and bottom.

Horizontal red and green lines separated at top.

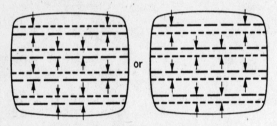

Horizontal red and green lines separated overall (mainly at top and bottom).

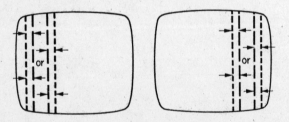

Vertical red and green lines separated at left or at right.

Fig. 71. Dynamic convergence errors.

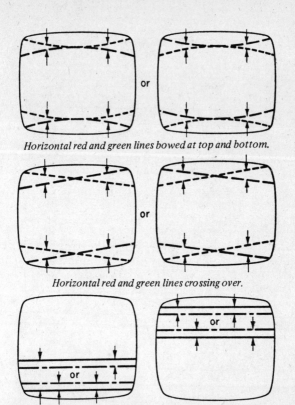

Horizontal red and green lines bowed at top and bottom.

Horizontal red and green lines crossing over.

Horizontal blue and yellow lines separated at bottom (left) and top (right).

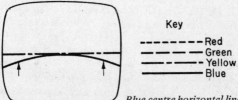

Key

------- Red
– – – Green
–·–·– Yellow
——— Blue

Blue centre horizontal line drooping.

123

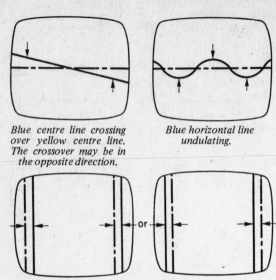

Blue centre line crossing over yellow centre line. The crossover may be in the opposite direction.

Blue horizontal line undulating.

Vertical blue and yellow lines separated at sides.

Fig. 71 (continued)

then check the raster dimensions and adjust as necessary with the timebase controls. The circuits will have been carefully set up during manufacture so that adjustment on installation should not be too extensive.

Although the effects on the cross-hatch pattern of this fairly standard set of controls are much the same from model to model, the recommended order of adjustment varies considerably so that the official service manual should be consulted before dynamic convergence adjustments are undertaken.

Finally note that perfect convergence over the entire picture cannot be achieved—the residual errors will be greatest at the corners of the picture. Concentrate on obtaining correct convergence along the centre vertical and horizontal lines of the cross-hatch pattern. Some slight compromise readjustment

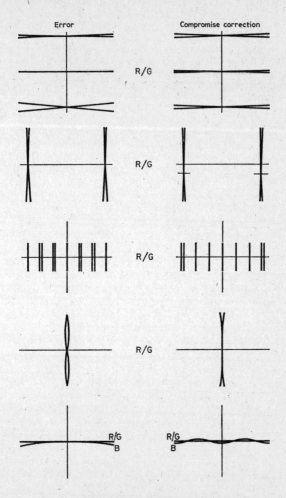

Error   Compromise correction

R/G

R/G

R/G

R/G

R/G   R/G
B   B

Fig. 72. Permissible compromise adjustments to spread residual convergence errors.

125

is finally permissible, along the lines shown in Fig. 72, to spread residual convergence errors.

Are additional problems presented by the use of 110° picture tubes with regard to raster registration and convergence?

It is true that the wider scanning angle calls for greater detail in the techniques to minimise pincushion distortion and to optimise the convergence over the entire area of screen, but modern circuits and scanning components are significantly helping to cut the complications to a bare minimum. Indeed, some of the latest 110° receivers are no more difficult to adjust convergence-wise than their earlier counterparts—it has been suggested that some are less difficult, eased by the lack of the dual-standard requirement of earlier days.

Various circuits have been evolved for pincushion correction and convergence, and some, it must be admitted, are quite complicated. However, at the time of writing a new breed of 110° receiver is emerging with relatively less complicated circuits. Starting with pincushion correction, this is achieved in the north/south (top/bottom) sense by a transductor similar to that employed in 90° receivers (see pages 60 and 117) whereby a small current at line frequency is superimposed on the field scanning current such that the scanning current is increased at the centre of each line at the top and bottom sections of the picture. Further correction is provided by the introduction of a second harmonic component to help straighten the lines at the top of the picture.

As with 90° receivers, side (east/west) pincushion distortion is corrected by the application of a parabolic current at field rate to the line scanning coils, this having the effect of reducing the line amplitude at the start and finish of each field scan. Unlike 90° receivers, however, 110° models often have an active east/west circuit, consisting of an east/west pincushion amplifier, which yields the greater range of correction required by the larger scanning angle. In one recent receiver a three-stage (transistorised) amplifier is adopted which receives a low amplitude parabolic voltage, fed via a preset for east/west amplitude control, and a sawtooth waveform, also

126

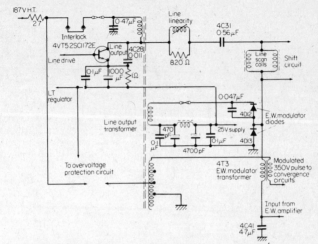

Fig. 73. Basic circuit of pincushion distortion correction (east/west sense) of RBM Z179 chassis, showing also basic line output stage.

applied via a preset, which enables the parabola to be tilted for the correction of keystone distortion. The output of the amplifier then feeds a signal, via a winding on the east/west modulation transformer, in series with the line scanning coils, as shown in Fig. 73 (which also reveals the basic line output stage). This circuit pertains to the Rank-Bush-Murphy Z179 110° chassis. Another interesting aspect of the east/west amplifier is that by varying its bias the d.c. level at the collector of the output transistor is altered, which in turn varies the line scan amplitude linearly across the field, so the bias control serves as a linear width control.

The convergence circuits of the same receiver are given in Fig. 74. Starting with the horizontal section at top left, both red/green and blue circuits are shunt fed from the line output transformer, and for increased deflection sensitivity at the corners of the picture the voltage waveform is modulated by the field scan, this by 4T3 (Fig. 73). Double integration is employed, first by 5T1 and 5T2 and second by 7L9, 7L10 and

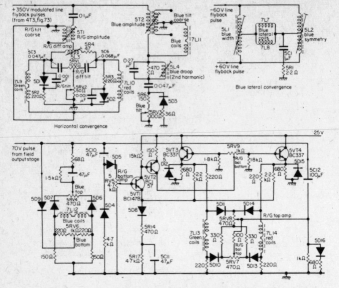

Fig. 74. Convergence circuits of the RBM Z197 110° chassis.

7L11. Tilt is provided by 5R2/5R3 (red/green) and by 5R6/5RV3 (blue). Diodes 5D1/5D2 clamp the centre of the display to avoid static misconvergence while dynamic convergence is under adjustment. Red/green amplitude is controlled by 5L3, while 5R4/5C3/5RV1/5C6 (damped resonant circuit) assist with horizontal waveform shaping.

The blue lateral coils are 7L7/7L8 (top-right circuit section), and the required sawtooth and parabolic currents of either polarity are yielded by the integration of ±60V line pulses. Integration by 5L1 feeds sawtooth current to the coils while 5L2 (with 5C1/5R1) supplies also parabolic current to the coils.

Vertical convergence utilises the bottom section of the circuit, which is partly active (i.e. uses transistors). For red/green vertical convergence field-derived signal is introduced in

128

two ways. For the first half of the scan it goes by way of 5C15 direct to the matrixing network and for the second half via the clipping circuit 5C10/5D4/5D5 to 5VT1 base. 5D8/5R14/etc. in the emitter circuit of this transistor constitute a shaping circuit. Stage 5VT2 is an inverting stage which drives a pair of symmetrical emitter-followers 5VT3/3VT4.

5D12/5D15 bypass field flyback pulses to chassis. Since there is some interaction between the voltage drive to 5VT3/5VT4 bases as adjusted by 5RV9 and the red/green "top" controls, since these form part of the emitter loads of the transistors, the vertical convergence is adjusted first at the bottom and afterwards at the top. The adjustment of the "top" controls, however, does not affect the setting of the "bottom" control.

Blue vertical convergence may also require current of either polarity, and this is provided by driving each end of 7L12. As before, bottom adjustment tends to affect top convergence, so the plan is to adjust the bottom of the picture first and then the top.

Mention was made in Section 2 about a thyristor line output circuit, could you give brief details of this?

This circuit was developed by ITT for use with their thin-neck 110° tubes, and is shown in simplified form in Fig. 75. Briefly, the circuit works like this. At line scan centre C4 is positively charged with respect to chassis and current from the drive signal has just been communicated to the thyristor (SCR2) gate, thereby triggering the device on. This causes current to flow constantly through the scanning coils, via C4 and SCR2, thereby giving a linear scan over the second half of a line. At the end of the scan a line pulse arrives to trigger SCR1, but just before this happens there is a current build-up through L1/C3 from the scanning half of the circuit which produces a kind of overshoot, called a "commutating pulse", the effect of which momentarily reverse-biases the SCR2, the reverse current flowing to chassis through D2. After the mutating pulse SCR2 and D2 switch off, SCR1 then conducting while the energy stored in the scanning coils flows out

129

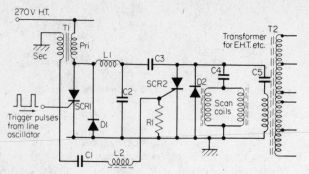

Fig. 75. Simplified thyristor line output stage designed by ITT for use with their thin-neck 100° tubes.

through C3/L1 charging C3. When the energy has been dissipated half the flyback has occurred SCR1 switches off, D1 then conducting, the charge in C3 providing the current required for the remainder of the flyback. At the conclusion of the flyback C3 is discharged, but now the scanning coils contain significant energy and this is dissipated to chassis through D2, the current through the scanning coils due to this then giving the first half of a linear line scan. After the first half of a line scan all the energy in the scanning coils has been exhausted, but C4 is charged, so the whole sequence starts all over again. The pulse obtained via C1/L2 ensures that SCR2 conducts at the correct time, D2 then switching off. The drive from the line oscillator, applied to SCR1 gate, initiates the flyback action.

INDEX

131